Plant Anatomy

Plant Anatomy

Melody Sawyer

States Academic Press,
109 South 5th Street,
Brooklyn, NY 11249, USA

Visit us on the World Wide Web at:
www.statesacademicpress.com

This book contains information obtained from authentic and highly regarded sources. All chapters are published with permission under the Creative Commons Attribution Share Alike License or equivalent. A wide variety of references are listed. Permissions and sources are indicated; for detailed attributions, please refer to the permissions page. Reasonable efforts have been made to publish reliable data and information, but the authors, editors and publisher cannot assume any responsibility for the validity of all materials or the consequences of their use.

ISBN: 978-1-63989-418-5 (Hardback)

Trademark Notice: Registered trademark of products or corporate names are used only for explanation and identification without intent to infringe.

Cataloging-in-Publication Data

Plant anatomy / Melody Sawyer.
     p. cm.
Includes bibliographical references and index.
ISBN 978-1-63989-418-5
1. Plant anatomy. 2. Botany. 3. Anatomy. I. Sawyer, Melody.
QK641 .P53 2022
581.4--dc23

TABLE OF CONTENTS

This book aims to help a broader range of students by exploring a wide variety of significant topics related to this discipline. It will help students in achieving a higher level of understanding of the subject and excel in their respective fields. This book would not have been possible without the unwavered support of my senior professors who took out the time to provide me feedback and help me with the process. I would also like to thank my family for their patience and support.

The branch of biology that studies the internal structure of plants is referred to as plant anatomy. This field includes microscopy of plants and the sectioning of its tissues. The study of plant anatomy uses systems approach which is based on the plant's activities. These activities are flowering, seed development, pollination and nutrient transport. The discipline is divided further into structural categories. Flower anatomy studies gynoecium, calyx, corolla and androecium. Leaf anatomy focuses on the study of epidermis, palisade cells and stomata. Stem anatomy studies stem structure, vascular tissues, shoot apex and buds. Root anatomy includes the study of root structure, its tip and endodermis. Fruit and seed anatomy refers to the study of structure of the ovule, seed, accessory fruit and pericarp. This book is a compilation of chapters that discuss the most vital concepts in the field of plant anatomy. Some of the diverse topics covered herein address the varied branches that fall under this category. Through this book, we attempt to further enlighten the readers about the new concepts in this field.

A brief overview of the book contents is provided below:

Chapter – Introduction

A plant mainly consists of three major vegetative organs namely root, stem and leaf, as well as reproductive parts like flower, fruits and seeds. Plant anatomy is the field of biology that deals with the study of internal structure of plants. This is an introductory chapter which will briefly introduce about plant anatomy.

Chapter – Plant Cells and Tissues

The eukaryotic cells present in green plants that belong to the kingdom Plantae are referred to as plant cells. Plants have three tissue types – dermal, ground and vascular tissues. This chapter has been carefully written to provide an easy understanding of all the aspects related to plant cells and tissues.

Chapter – Stem and Root Anatomy

The stem is the ascending part of the plant which is divided into nodes and internodes. It bears leaves, branches and flowers. Parts of a root include the primary root, lateral roots, the apical meristem, a root cap, and root hair. The topics elaborated in this chapter will help in gaining a better perspective of stem and root anatomy.

Chapter – Flower and Leaf Anatomy

Flower is the reproductive part in flowering plants. It mainly consists of four parts which are sepals, petals, stamens and carpels. The leaf is the photosynthetic organ of the plant. It comprises blade, petiole and leaflets. The chapter discusses the anatomy of flower and leaf in detail.

Chapter – Fruit and Seed Anatomy

Fruits are the mature ovaries of one or more flowers. Fruit anatomy is the internal structure of fruits. Seeds are matured and fertilized ovules. It consists of three components – embryo, endosperm and seed coat. This chapter closely examines fruit and seed anatomy to provide an extensive understanding of the subject.

Melody Sawyer

Introduction 1

- **Plant**
- **Plant Anatomy**

A plant mainly consists of three major vegetative organs namely root, stem and leaf, as well as reproductive parts like flower, fruits and seeds. Plant anatomy is the field of biology that deals with the study of internal structure of plants. This is an introductory chapter which will briefly introduce about plant anatomy.

Plant

Plant, (Kingdom Plantae) is any multicellular eukaryotic life-form characterized by: (1) Photosynthetic nutrition (a characteristic possessed by all plants except some parasitic plants and underground orchids), in which chemical energy is produced from water, minerals, and carbon dioxide with the aid of pigments and the radiant energy of the Sun, (2) Essentially unlimited growth at localized regions, (3) Cells that contain cellulose in their walls and are therefore to some extent rigid, (4) The absence of organs of locomotion, resulting in a more or less stationary existence, (5) The absence of nervous systems, and (6) Life histories that show an alteration of haploid and diploid generations, with the dominance of one over the other being taxonomically significant.

Weeping willow (Salix babylonica).

Highveld grassland.

Plants range in size from diminutive duckweeds only a few millimetres in length to the giant sequoias of California that reach 90 metres (300 feet) or more in height. There are an estimated 390,900 different species of plants known to science, and new species are continually being described, particularly from previously unexplored tropical areas of the world. Plants evolved from aquatic ancestors and have subsequently migrated over the entire surface of Earth, inhabiting tropical, Arctic, desert, and Alpine regions. Some plants have returned to an aquatic habitat in either fresh or salt water.

Duckweed (Lemna minor).

Plants play a vital role in the maintenance of life on Earth. All energy used by living organisms depends on the complex process of photosynthesis, which is mostly carried out by green plants. Radiant energy from the Sun is transformed into organic chemical energy in the form of sugars through the fundamental series of chemical reactions constituting photosynthesis. In nature all food chains begin with photosynthetic autotrophs (primary producers), including green plants and algae. Primary producers, represented by trees, shrubs, and herbs, are a prolific source of energy in the form of carbohydrates (sugars) stored in the leaves. These carbohydrates, produced in photosynthesis, are broken down in a process called respiration; the smaller units of the sugar molecule and its products fuel numerous metabolic processes. Various parts of the plant (e.g., leaves) are the energy sources that support animal life in different community habitats. A by-product of photosynthesis, oxygen, is essential to animals.

Sequoia tree.

The daily existence of human beings is also directly influenced by plants. Plants furnish food and flavourings; raw materials for industry, such as wood, resins, oils, and rubber; fibres for the manufacture of fabrics and cordage; medicines; insecticides; and fuels. More than half of Earth's population relies on the grasses rice, corn (maize), and wheat as their primary source of food. Apart from their commercial and aesthetic value, plants conserve other natural resources by protecting soils from erosion, by controlling water levels and quality, and by producing a favourable atmosphere.

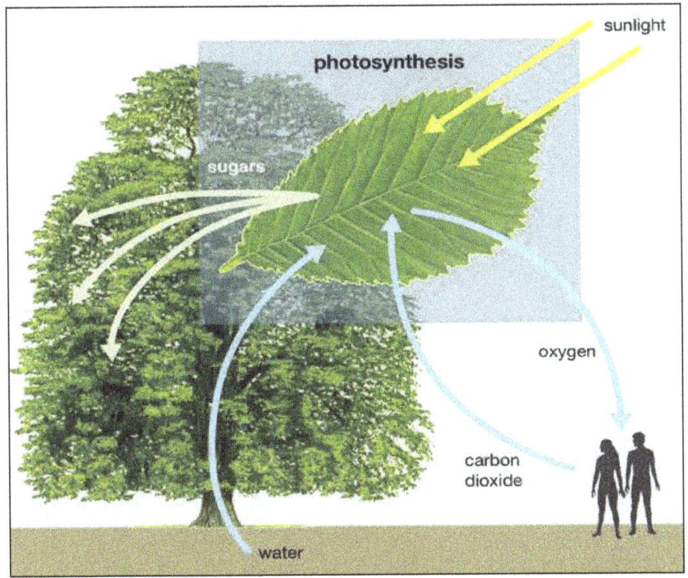

Diagram of photosynthesis showing how water, light, and carbon dioxide are absorbed by a plant to produce oxygen, sugars, and more carbon dioxide.

Definition of the Kingdom

The kingdom Plantae includes organisms that range in size from tiny mosses to giant trees. Despite this enormous variation, all plants are multicellular and eukaryotic (i.e., each cell possesses a membrane-bound nucleus that contains the chromosomes). They generally possess pigments (chlorophylls a and b and carotenoids), which play a central role in converting the energy of sunlight into chemical energy by means of photosynthesis. Most plants, therefore, are independent in their nutritional needs (autotrophic) and store their excess food in the form of macromolecules of starch. The relatively few plants that are not autotrophic have lost pigments and are dependent on other organisms for nutrients. Although plants are nonmotile organisms, some produce motile cells (gametes) propelled by whiplike flagella. Plant cells are surrounded by a more or less rigid cell wall composed of the carbohydrate cellulose, and adjacent cells are interconnected by microscopic strands of cytoplasm called plasmodesmata, which traverse the cell walls. Many plants have the capacity for unlimited growth at localized regions of cell division, called meristems. Plants, unlike animals, can use inorganic forms of the element nitrogen (N), such as nitrate and ammonia—which are made available to plants through the activities of microorganisms or through the industrial production of fertilizers—and the element sulfur (S); thus, they do not require an external source of protein (in which nitrogen is a major constituent) to survive.

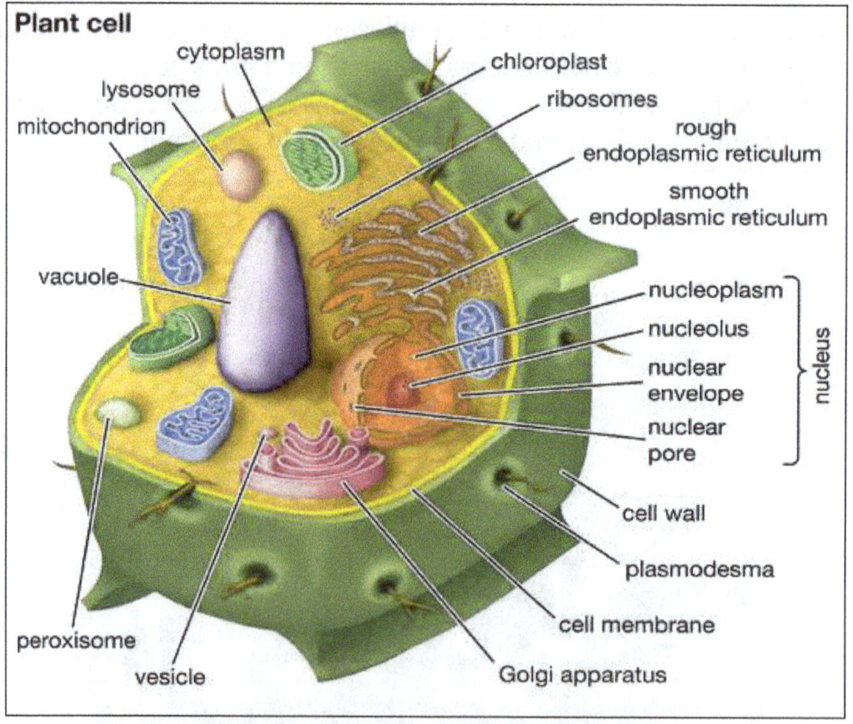

Cutaway drawing of a plant cell, showing the cell wall and internal organelles.

The life histories of plants include two phases, or generations, one of which is diploid (the nuclei of the cells contain two sets of chromosomes), whereas the other is haploid (with one set of chromosomes). The diploid generation is known as the sporophyte, which literally means spore-producing plant. The haploid generation, called the gametophyte, produces the sex cells, or gametes. The complete life cycle of a plant thus involves an alternation of generations. The sporophyte and gametophyte generations of plants are structurally quite dissimilar.

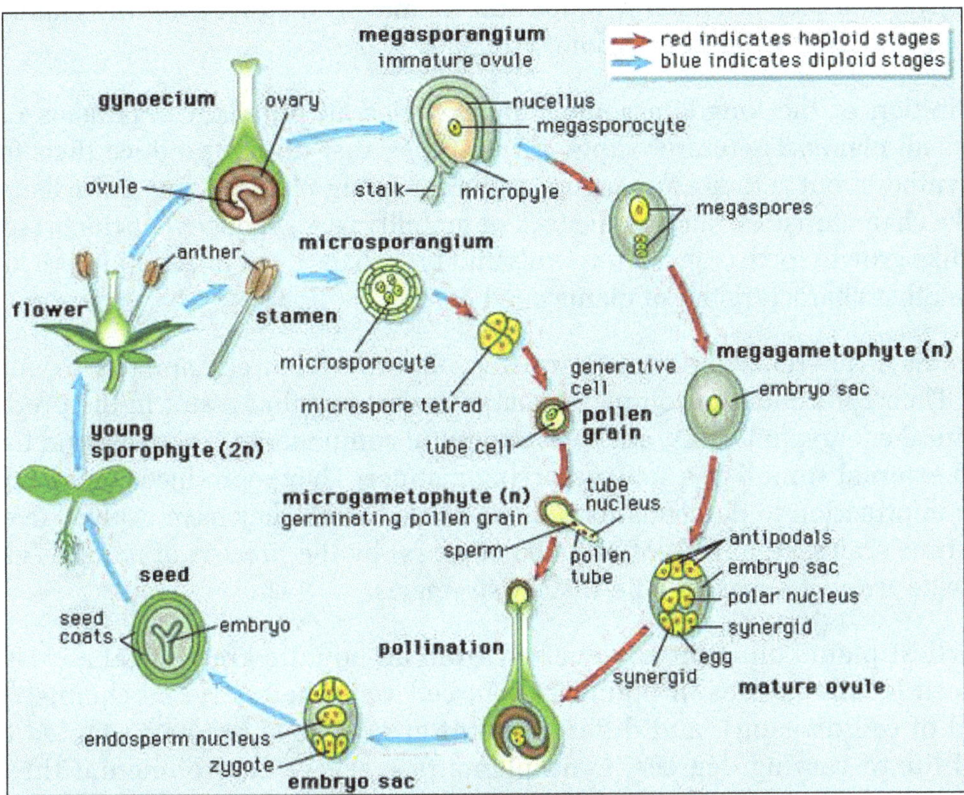

Life cycle of a typical angiosperm.

The angiosperm life cycle consists of a sporophyte phase and a gametophyte phase. The cells of a sporophyte body have a full complement of chromosomes (i.e., the cells are diploid, or 2n); the sporophyte is the typical plant body that one sees when one looks at an angiosperm. The gametophyte arises when cells of the sporophyte, in preparation for reproduction, undergo meiotic division and produce reproductive cells that have only half the number of chromosomes (i.e., haploid, or n). A two-celled microgametophyte (called a pollen grain) germinates into a pollen tube and through division produces the haploid sperm. An eight-celled megagametophyte (called the embryo sac) produces the egg. Fertilization occurs with the fusion of a sperm with an egg to produce a zygote, which eventually develops into an embryo. After fertilization, the ovule develops into a seed, and the ovary develops into a fruit.

The concept of what constitutes a plant has undergone significant change over time.

For example, at one time the photosynthetic aquatic organisms commonly referred to as algae were considered members of the plant kingdom. The various major algal groups, such as the green algae, brown algae, and red algae, are now placed in the kingdom Protista because they lack one or more of the features that are characteristic of plants. The organisms known as fungi also were once considered to be plants because they reproduce by spores and possess a cell wall. The fungi, however, uniformly lack chlorophyll, and they are heterotrophic and chemically distinct from the plants; thus, they are placed in a separate kingdom, Fungi.

No definition of the kingdom completely excludes all nonplant organisms or even includes all plants. There are plants, for example, that do not produce their food by photosynthesis but rather are parasitic on other living plants. Some animals possess plantlike characteristics, such as the lack of mobility (e.g., sponges) or the presence of a plantlike growth form (e.g., some corals and bryozoans), but in general such animals lack the other characteristics of plants cited here.

Despite such differences, plants share the following features common to all living things. Their cells undergo complex metabolic reactions that result in the production of chemical energy, nutrients, and new structural components. They respond to internal and external stimuli in a self-preserving manner. They reproduce by passing their genetic information to descendants that resemble them. They have evolved over geological time scales (hundreds of millions of years) by the process of natural selection into a wide array of forms and life-history strategies.

The earliest plants undoubtedly evolved from an aquatic green algal ancestor (as evidenced by similarities in pigmentation, cell-wall chemistry, biochemistry, and method of cell division), and different plant groups have become adapted to terrestrial life to varying degrees. Land plants face severe environmental threats or difficulties, such as desiccation, drastic changes in temperature, support, nutrient availability to each of the cells of the plant, regulation of gas exchange between the plant and the atmosphere, and successful reproduction. Thus, many adaptations to land existence have evolved in the plant kingdom and are reflected among the different major plant groups. An example is the development of a waxy covering (the cuticle) that covers the plant body, preventing excess water loss. Specialized tissues and cells (vascular tissue) enabled early land plants to absorb and transport water and nutrients to distant parts of the body more effectively and, eventually, to develop a more complex body composed of organs called stems, leaves, and roots. The evolution and incorporation of the substance lignin into the cell walls of plants provided strength and support. Details of the life history are often a reflection of a plant's adaptation to a terrestrial mode of life and may characterize a particular group; for example, the most highly evolved plants reproduce by means of seeds, and, in the most advanced of all plants (angiosperms), a reproductive organ called a flower is formed.

era	period	time (millions of years ago)	algae	bryophytes (mosses)	seedless vascular plants (ferns, horsetails)	ginkgoes	conifers	cycads	angiosperms (flowering plants)
Cenozoic	Quaternary	2.6							
	Neogene	23							
	Paleogene	66							
Mesozoic	Cretaceous	145							flowering plants appear about 130 million years ago
	Jurassic	201							
	Triassic	252							
Paleozoic	Permian	299							
	Carboniferous	359				first seed plants appear about 383–359 million years ago			
	Devonian	419			early vascular plants appear about 410 million years ago				
	Silurian	444							
	Ordovician	485							

Significant events in plant evolution.

Nonvascular Plants

Informally known as bryophytes, nonvascular plants lack specialized vascular tissue (xylem and phloem) for internal water and food conduction and support. They also do not possess true roots, stems, or leaves. Some larger mosses, however, contain a central core of elongated thick-walled cells called hydroids that are involved in water conduction and that have been compared to the xylem elements of other plants. Bryophytes are second in diversity only to the flowering plants (angiosperms) and are generally regarded as composed of three divisions: Bryophtya (the mosses), Marchantiophyta (the liverworts), and Anthocerotophyta (the hornworts).

Red carpet moss (Bryoerythrophyllum columbianum).

Because bryophytes generally lack conducting cells and a well-developed cuticle that would limit dehydration, they depend on their immediate surroundings for an adequate supply of moisture. As a result, most bryophytes live in moist or wet shady locations, growing on rocks, trees, and soil. Some, however, have become adapted to totally aquatic habitats; others have become adapted to alternately wet and dry environments by growing during wet periods and becoming dormant during dry intervals. Although bryophytes are widely distributed, occurring in practically all parts of the world, none are found in salt water. Ecologically, some mosses are considered pioneer plants because they can invade bare areas.

Bryophytes are typically land plants but seldom attain a height of more than a few centimetres. They possess the photosynthetic pigment chlorophyll (both a and b forms) and carotenoids in cell organelles called chloroplasts. The life histories of these plants show a well-defined alternation of generations, with the independent and free-living gametophyte as the dominant photosynthetic phase in the life cycle. (This is in contrast to the vascular plants, in which the dominant photosynthetic phase is the sporophyte.). The sporophyte generation develops from, and is almost entirely parasitic on, the gametophyte. The gametophyte produces multicellular sex organs (gametangia). Female gametangia are called archegonia; male gametangia, antheridia. At maturity, archegonia each contain one egg, and antheridia produce many sperm cells. Because the egg is retained and fertilized within the archegonium, the early stages of the developing sporophyte are protected and nourished by the gametophytic tissue. The young undifferentiated sporophyte is called an embryo. Although bryophytes have become adapted to life on land, an apparent vestige of their aquatic ancestry is that the motile (flagellated) sperm depend on water to allow gamete transport and fertilization.

Liverwort archegonia.

Bryophytes are widely believed to have evolved from complex green algae that invaded land around 500 million years ago. Bryophytes share some traits with green algae, such as motile sperm, similar photosynthetic pigments, and the general absence of vascular tissue. However, bryophytes have multicellular reproductive structures, whereas those of green algae are unicellular, and bryophytes are mostly terrestrial and have complex plant bodies, whereas the green algae are primarily aquatic and have less-complex forms.

Representative Members

Division Bryophyta

Moss is a term erroneously applied to many different plants (Spanish moss, a flowering plant; Irish moss, a red alga; pond moss, filamentous algae; and reindeer moss, a lichen). True mosses are classified as the division Bryophyta.

Peat moss (*Sphagnum flexuosum*).

The moss gametophyte possesses leaflike structures (phyllids) that usually are a single cell layer thick, have a costa (midrib), and are spirally arranged on a stemlike axis (caulid). The moss gametophyte is an independent plant and is the familiar, erect "leafy" shoot. Multicellular rhizoids anchor the gametophyte to the substrate. The sporophyte plant develops from the tip of the fertile leafy shoot. After repeated cell divisions, the young sporophyte (embryo) transforms into a mature sporophyte consisting of foot, elongate seta, and capsule. The capsule is often covered by a calyptra, which is the enlarged remains of the archegonium. The capsule is capped by an operculum (lid), which falls off, exposing a ring of teeth (the peristome) that regulates the dispersal of spores.

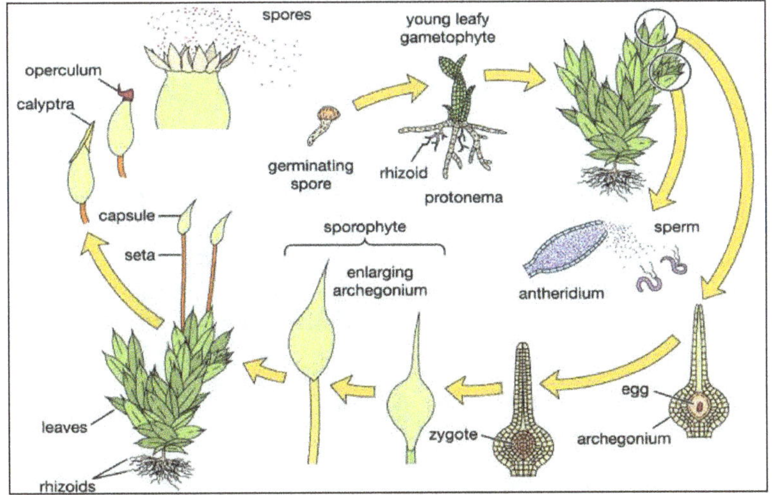

Life cycle of moss.

Division Marchantiophyta

Liverworts, the second major division of nonvascular plants, are found in the same types of habitat as mosses, and species of the two classes are often intermingled on the same site. The curious name liverwort is a relic of the medieval belief in the "doctrine of signatures," which held that the external form of a plant provided a clue to which diseased body organ could be cured by a preparation made from that particular plant. There are two types of liverworts (also called hepatics) based on reproductive features and thallus structure. The more numerous "leafy" liverworts superficially resemble mosses, but most notably differ in having lobed or divided leaves that are without a midrib and are positioned in three rows. Thalloid (thallose) liverworts have a ribbonlike, or strap-shaped, body that grows flat on the ground. They have a high degree of internal structural differentiation into photosynthetic and storage zones. Liverwort gametophytes have unicellular rhizoids. Liverworts have an alternation of generations similar to that of mosses, and, as with mosses, the gametophyte generation is dominant. The sporophytes, however, are not microscopic and are often borne on specialized structures. They sometimes resemble small umbrellas and are called antheridiophores and archegoniophores.

Thalloid of the liverwort Marchantia with gemma cups.

Division Anthocerotophyta

Hornwort (Dendroceros).

The third division of bryophytes comprises the hornworts, a minor group numbering fewer than 100 species. The gametophyte is a small ribbonlike thallus that resembles a thallose liverwort. The name hornwort is derived from the unique slender, upright sporophytes, which are about 3–4 cm (1.2–1.6 inches) long at maturity and dehisce longitudinally into two valves that twist in response to changing humidity, thereby releasing spores in small numbers over a fairly long period of time.

Vascular Plants

Vascular plants (tracheophytes) differ from the nonvascular bryophytes in that they possess specialized supporting and water-conducting tissue, called xylem, and food-conducting tissue, called phloem. The xylem is composed of nonliving cells (tracheids and vessel elements) that are stiffened by the presence of lignin, a hardening substance that reinforces the cellulose cell wall. The living sieve elements that comprise the phloem are not lignified. Xylem and phloem are collectively called vascular tissue and form a central column (stele) through the plant axis. The ferns, gymnosperms, and flowering plants are all vascular plants. Because they possess vascular tissues, these plants have true stems, leaves, and roots. Before the development of vascular tissues, the only plants of considerable size existed in aquatic environments where support and water conduction were not necessary. A second major difference between the vascular plants and bryophytes is that the larger, more conspicuous generation among vascular plants is the sporophytic phase of the life cycle.

Tree fern (Cyathea medullaris).

The vegetative body of vascular plants is adapted to terrestrial life in various ways. In addition to vascular tissue, the aerial body is covered with a well-developed waxy layer (cuticle) that decreases water loss. Gases are exchanged through numerous pores (stomata) in the outer cell layer (epidermis). The root system is involved in the uptake from the soil of water and minerals that are used by the root system as well as the stem and leaves. Roots also anchor the plant and store food. The stem conducts water and minerals absorbed by the root system upward to various parts of the stem and leaves;

stems also conduct carbohydrates manufactured through the process of photosynthesis from the leaves to various parts of the stem and root system. Leaves are supported by the stem and are oriented in a manner conducive to maximizing the amount of leaf area involved in trapping sunlight for use in photosynthesis.

Spring flowering of bluebells (*Hyacinthoides nonscripta*) covering the floor of a deciduous forest of beech (*Fagus sylvatica*) and oak (*Quercus*).

Modifications of roots, stems, and leaves have enabled species of vascular plants to survive in a variety of habitats encompassing diverse and even extreme environmental conditions. The ability of vascular plants to flourish in so many different habitats is a key factor in their having become the dominant group of terrestrial plants.

The vascular plants are divisible into the nonseed plants (lower vascular plants, or cryptogams) and those that reproduce by seeds (higher vascular plants, or phanerogams). The ferns (Polypodiopsida) are a group of the lower vascular plants; other groups include the whisk ferns (Psilotophyta), club and spike mosses (Lycophyta), and horsetails (Sphenophyta, or Arthrophyta). Collectively, the latter four groups are sometimes referred to as pteridophytes, because each reproduces by spores liberated from dehiscent sporangia (free sporing). Although the lower vascular plants have adapted to terrestrial life, they are similar to bryophytes in that, as an apparent vestige of their aquatic ancestry, all produce motile (flagellated) male gametes (antherozoids, or sperm) and must rely on water for fertilization to take place.

Nonseed Plants

Division Lycophyta

This division is represented by four or more living genera, with the principal genera being Lycopodium (club mosses), Selaginella (spike mosses), and Isoetes (quillworts). Extant members of Lycophyta occur in both temperate and tropical regions and represent the survivors of a group of vascular plants that was extremely diverse and numerous. As a group, the lycopods were prominent in the great coal-forming swamp forests of the Carboniferous

Period (358.9 million to 298.9 million years ago). Although all living lycopods are small herbaceous plants, some extinct types were large trees. Lycopods are differentiated into stem, root, and leaf (microphylls). Sporangia are positioned on the upper (adaxial) surface of the leaf (sporophyll). Some species form distinct cones or strobili, whereas others do not.

Spike moss (Selaginella).

Ferns

Ferns are a diverse group of plants that are unranked at the division level in some taxonomies. Formerly, the group was designated as division Pteridophtya, but their phylogenetic relationships remain unresolved. Although they have a worldwide distribution, ferns are more common in tropical and subtropical regions. They range in size and complexity from small floating aquatic plants less than 2 cm (0.8 inch) long to tall tree ferns 20 metres (65 feet) high. Tropical tree ferns possess erect columnar trunks and large compound (divided) leaves more than 5 metres (about 16 feet) long. As a group, ferns are either terrestrial or epiphytic (growing upon another plant). Fern stems never become woody (composed of secondary tissue containing lignin), because all tissues of the plant body originate at the stem apex.

Shield fern (Dryopteris dilatata).

Class Polypodiopsida

Ferns of the class Polypodiopsida typically possess a rhizome (horizontal stem) that grows partially underground; the deeply divided fronds (leaves) and the roots grow out of the rhizome. Fronds are characteristically coiled in the bud (fiddleheads) and uncurl in a type of leaf development called circinate vernation. Fern leaves are either whole or variously divided. The leaf types are differentiated into rachis (axis of a compound leaf), pinnae (primary divisions), and pinnules (ultimate segments of a pinna). Fern leaves often have prominent epidermal hairs and large chaffy scales. Venation of fern leaves is usually open dichotomous (forking into two equal parts).

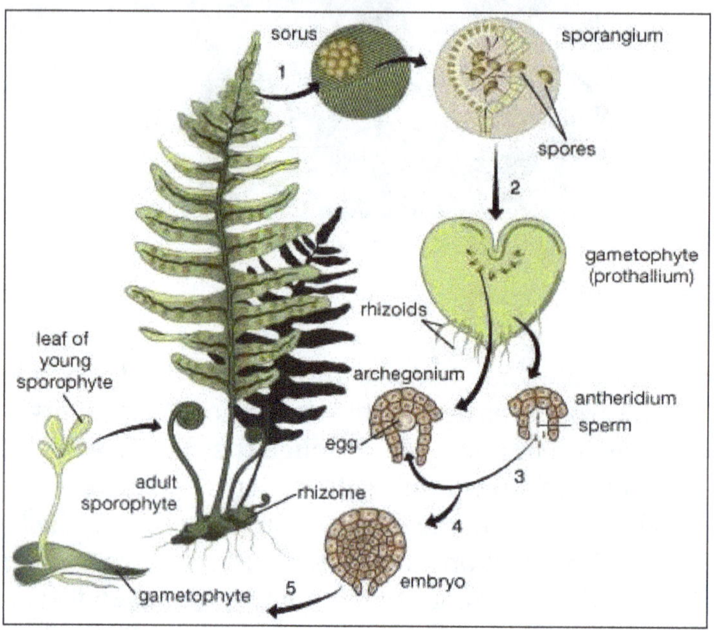

The life cycle of the fern.

In the above figure: (1) Clusters (sori) of sporangia (spore cases) grow on the under-surface of mature fern leaves. (2) Released from its spore case, the haploid spore is carried to the ground, where it germinates into a tiny, usually heart-shaped, gameto-phyte (gamete-producing structure), anchored to the ground by rhizoids (rootlike projections). (3) Under moist conditions, mature sperm are released from the antheridia and swim to the egg-producing archegonia that have formed on the gametophyte's lower surface. (4) When fertilization occurs, a zygote forms and develops into an embryo within the archegonium. (5) The embryo eventually grows larger than the gametophyte and becomes a sporophyte.

Each frond is a potential sporophyll (spore-bearing leaf) and as such can bear structures that are associated with reproduction. When growth conditions are favourable, a series of brown patches appear on the undersurface of the sporophylls. Each one of the patches (called a sorus) is composed of many sporangia, or spore cases, which are joined by a stalk to the sporophyll. The spore case is flattened, with a layer of sterile, or nonfertile, cells surrounding the spore mother cells. Each spore mother cell divides by reduction division (meiosis) to produce haploid spores, which are shed in a way characteristic to the ferns.

Each fern spore has the potential to grow into a green heart-shaped independent gametophyte plant (prothallus) capable of photosynthesis. In contrast to bryophytes, in which the sporophyte is nutritionally dependent on the gametophyte during its entire existence, the fern sporophyte is dependent on the gametophyte for nutrition only during the early phase of its development; thereafter, the fern sporophyte is free-living. In some ferns the sexes are separate, meaning a gametophyte will bear only male or female sex organs. Other species have gametophytes bearing both sex organs. Features important in the identification of ferns include such aspects of the mature sporophyte plant as differences in the stem, frond, sporophyll, sporangium, and position of the sporangium and the absence or presence, as well as the shape, of the indusium (a membranous outgrowth of the leaf) covering the sporangia.

Class Psilotopsida

Psilotopsida (whisk ferns) is a class represented by two living genera (Psilotum and Tmesipteris) and several species that are restricted to the subtropics. This unusual group of small herbaceous plants is characterized by a leafless and rootless body possessing a stem that exhibits a primitive dichotomous type of branching: it forks into equal halves. The photosynthetic function is assumed by the stem, and the underground rhizome anchors the plant. The vascular tissue is organized into a poorly developed central cylinder in the stem.

Whisk fern (Psilotum nudum).

The family Ophioglossaceae, comprising four genera and some 80 species, is sometimes placed in the class Psilotopsida, though the taxonomy of the group is contentious.

Class Equisetopsida

Equisetopsida (also called horsetails and scouring rushes) is a class represented by a single living genus (Equisetum). It has a worldwide distribution but occurs in greater variety in the Northern Hemisphere. Like the lycopods, this group was a diverse and prominent group of vascular plants during the Carboniferous Period, when some genera attained great size in the coal-forming swamp forests. Known as sphenophytes, these plants are differentiated into stem, leaf (microphylls), and root. Green aerial stems have longitudinal ridges and furrows extending the length of the internodes, and stems are jointed (articulated). Surface cells are characteristically filled with silica. Branches, when they occur, are borne in whorls at the node, as are the scale leaves. Sporangia are borne in terminal strobili. Equisetopsida had its origin in the Devonian Period (419.2 million to 358.9 million years ago).

Giant horsetail of Europe (Equisetum telmateia).

Class Marattiopsida

Known as giant ferns, the class Marrattiopsida comprises a single extant family with four genera and some 150 species of large tropical and subtropical ferns with stout erect stems. The leaves (fronds) may be very large, some reaching 4.5 metres (15 feet) or more in length. The Marattiaceae generally are considered to be one of the most primitive families of ferns still living.

Seed Plants

Gymnosperms and angiosperms (flowering plants) share with ferns a dominant, independent sporophyte generation; the presence of vascular tissue; differentiation of the plant body into root, stem, and leaf derived from a bipolar embryo (having stem and root-growing apices); and similar photosynthetic pigments. Unlike ferns, however, the seed plants have stems that branch laterally and vascular tissue that is arranged in strands (bundles) around the pith (eustele). Among seed plants, as in ferns, the stem tissues that arise directly from the shoot apex are called primary tissues. Primary tissues contribute to the longitudinal growth of the stem, or primary growth. Secondary growth, resulting in an increase in the width of the axis, is produced by meristematic tissue between the primary xylem and phloem called vascular cambium. This meristem consists of a narrow zone of cells that form new secondary xylem (wood) and secondary phloem (secondary vascular tissues).

Two types of seed-bearing plants.

In the above figure:(Left) The Lawson cypress is an evergreen gymnosperm, or "naked seed" plant. It produces seeds in cones and bears needlelike leaves year-round. (Right) The English elm is a broad-leaved and deciduous angiosperm, or flowering plant. It produces seeds in fruits and drops its leaves in the autumn.

Major evolutionary advancements of these plants are demonstrated by the generally more complex plant body and by reproduction via seeds. Seeds represent an important evolutionary innovation within the plant kingdom. Each seed has an embryonic plant (sporophyte), food-storage tissue, and hardened protective covering (seed coat). The seed thus contains and protects the embryonic plant and, as the primary dispersal unit of the seed plants, represents a significant improvement over the spore, with its limited capacity for survival.

In comparing ferns and seed plants and their life histories, certain significant differences are seen. The gametophyte in seed plants has been reduced in size, usually consisting of a few to a dozen cells. Thus, it is no longer itself a plant body, as in the bryophytes and ferns. The gametophyte is not free-living but is embedded in the sporophyte and thus less vulnerable to environmental stress than the gametophytes of bryophytes and ferns. Finally, the spores of seed plants are male and female, as are the sporangia that contain them. The spores are not dispersed as in the bryophytes and ferns but develop into gametophytes within the sporangia. In the most advanced seed plants, the male gametes (sperm) are carried to the egg by a later extension of the pollen grain called the pollen tube. The advantage of this system is that the nonflagellated sperm are no longer dependent on water to reach the egg.

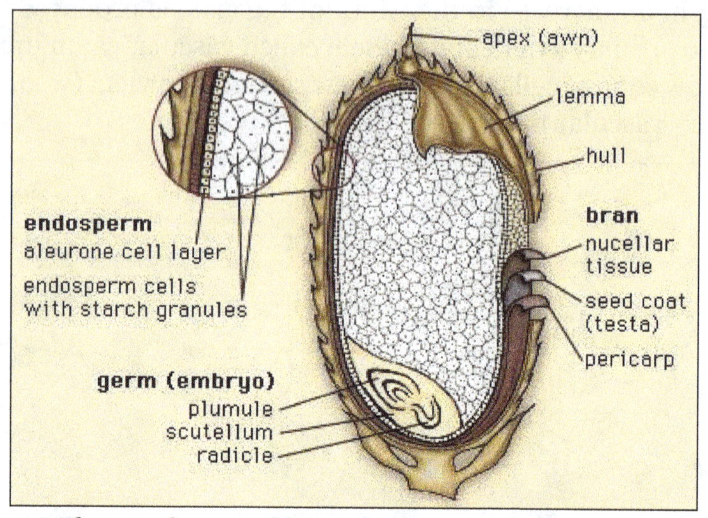

The outer layers and internal structures of a rice grain.

Another terrestrial adaptation of the seed plants not found in ferns is pollen dispersed by wind or animals. Pollen is a unit of genetic material as well as part of the seed-formation process. The dispersal of pollen by wind or animals, in addition to dispersal of seeds, promotes genetic recombination and distribution of the species over a wide geographic area.

Wind pollination in grasses: Yellow free-hanging anthers (pollen producers) and white feathery stigmas (pollen collectors) of meadow fescue (Festuca pratensis) provide maximum wind exposure.

Orange-tailed butterfly (Eurema proterpia) on an ash-coloured aster (Machaeranthera tephrodes). The upstanding yellow stamens are tipped with pollen, which brushes the body of the butterfly as it approaches the centre of the flat-topped aster to feed on the nectar.

Gymnosperms

The term gymnosperm ("naked seeds") represents four extant divisions of vascular plants whose ovules (seeds) are exposed on the surface of cone scales. The cone-bearing gymnosperms are among the largest and oldest living organisms in the world. They dominated the landscape about 200 million years ago. Today gymnosperms are of great economic value as major sources of lumber products, pulpwood, turpentine, and resins.

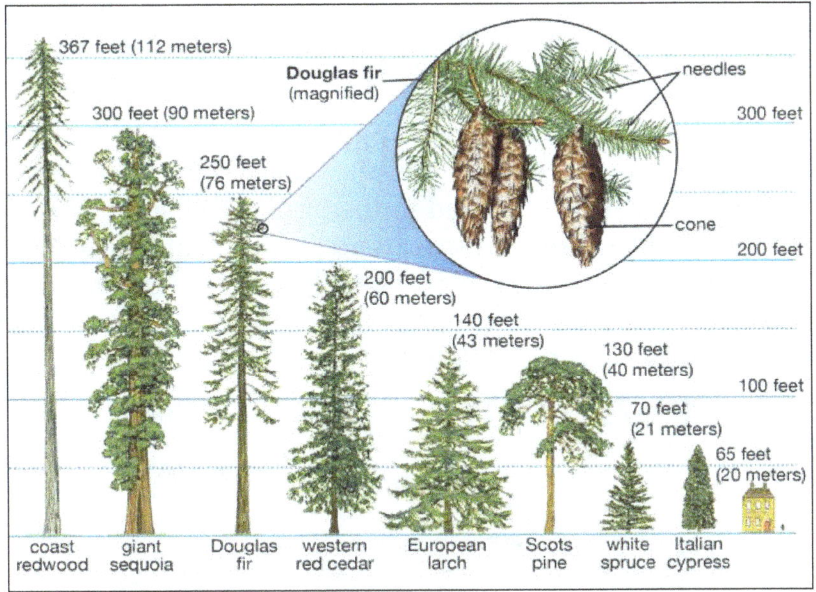

Conifer heights.

Conifer stems are composed of a woody axis containing primitive water- and mineral-conducting cells called tracheids. Tracheids are interconnected by passages called bordered pits. Leaves are often needlelike or scalelike and typically contain canals filled with resin. The leaves of pine are borne in bundles (fascicles), and the number of leaves per fascicle is an important distinguishing feature. Most gymnosperms are evergreen, but some, such as larch and bald cypress, are deciduous (the leaves fall after one growing season). The leaves of many gymnosperms have a thick cuticle and stomata below the leaf surface.

Cedar of Lebanon (Cedrus libani) showing (top) form and (bottom) leaves and cone.

The tree or shrub is the sporophyte generation. In conifers, the male and female sporangia are produced on separate structures called cones or strobili. Individual trees are typically monoecious (male and female cones are borne on the same tree). A cone is a modified shoot with a single axis, on which is borne a spirally arranged series of pollen- or ovule-bearing scales or bracts. The male cone, or microstrobilus, is usually smaller than the female cone (megastrobilus) and is essentially an aggregation of many small structures (microsporophylls) that encase the pollen in microsporangia.

The extant cycads (division Cycadophyta) are a group of ancient seed plants that are survivors of a complex that has existed since the Mesozoic Era (251.9 million to 66 million years ago). They are presently distributed in the tropics and subtropics of both hemispheres. Cycads are palmlike in general appearance, with an unbranched columnar trunk and a crown of large pinnately compound (divided) leaves. The sexes are always separate, resulting in male and female plants (i.e., cycads are dioecious). Most species produce conspicuous cones (strobili) on both male and female plants, and the seeds are very large.

Cycad (Cycas revoluta).

The ginkgophytes (division Ginkgophyta), although abundant, diverse, and widely distributed in the past, are represented now by a sole surviving species, Ginkgo biloba (maidenhair tree). The species was formerly restricted to southeastern China, but it is now likely extinct in the wild. The plant is commonly cultivated worldwide, however, and is particularly resistant to disease and air pollution. The ginkgo is multibranched, with stems that are differentiated into long shoots and dwarf (spur) shoots. A cluster of fan-shaped deciduous leaves with open dichotomous venation occurs at the end of each lateral spur shoot. Sexes are separate, and distinct cones are not produced. Female trees produce plumlike seeds with a fleshy outer layer and are noted for their foul smell when mature.

Leaves and fruit of the female ginkgo, or maidenhair tree (Ginkgo biloba).

The gnetophytes (division Gnetophyta) comprise a group of three unusual genera. Ephedra occurs as a shrub in dry regions in tropical and temperate North and South America and in Asia, from the Mediterranean Sea to China. Species of Gnetum occur as woody shrubs, vines, or broad-leaved trees and grow in moist tropical forests of South America, Africa, and Asia. Welwitschia, restricted to extreme deserts (less than 25 mm [1 inch] of rain per year) in a narrow belt about 1,000 km (600 miles) long in southwestern Africa, is an unusual plant composed of an enormous underground stem and a pair of long strap-shaped leaves that lie along the ground. The three genera differ from all other gymnosperms in possessing vessel elements (as compared with tracheids) in the xylem and in specializations in reproductive morphology. The gnetophytes have figured prominently in the theories about gymnospermous origins of the angiosperms.

Joint pine (Ephedra fragilis).

Gnetum gnemon.

Angiosperms

Approximately 130 million years ago, flowering plants (angiosperms) evolved from gymnosperms, although the identity of the specific gymnospermous ancestral group remains unresolved. The primary distinction between gymnosperms and angiosperms is that angiosperms reproduce by means of flowers. Flowers are modified shoots bearing a series of leaflike modified appendages and containing ovules (immature seeds) surrounded and protected by the female reproductive structure, the carpel or pistil. Along with other features, angiospermy, the enclosed condition of the seed, gave the flowering plants a competitive advantage and enabled them to come to dominate the extant flora. Flowering plants have also fully exploited the use of insects and other animals as agents of pollination (the transfer of pollen from male to female floral structures). In addition, the water-conducting cells and food-conducting tissue are more complex and efficient in flowering plants than in other land plants. Finally, flowering plants possess a specialized type of nutritive tissue in the seed, endosperm. Endosperm is the chief storage tissue in the seeds of grasses; hence, it is the primary source of nutrition in corn (maize), rice, wheat, and other cereals that have been utilized as major food sources by humans and other animals.

Sacred lotus (Nelumbo nucifera).

Talipot palm (Corypha umbraculifera) in bloom.

Classification of Angiosperms

Many of the flowering plants are commonly represented by two basic groups, the monocotyledons and the dicotyledons, distinguished by the number of embryonic seed leaves (cotyledons), number of flower parts, arrangement of vascular tissue in the stem, leaf venation, and manner of leaf attachment to the stem. However, one of the major changes in the understanding of the evolution of the angiosperms was the realization that the basic distinction among flowering plants is not between monocotyledon groups (monocots) and dicotyledon groups (dicots). Rather, plants thought of as being "typical dicots" have evolved from within another group that includes the more-basal dicots and the monocots together. This group of typical dicots is now known as the eudicots, and molecular-based evidence supports their having a single evolutionary lineage (monophyletic). Other angiosperm groups, such as the Magnoliids, do not fit the traditional paradigm of monocot and dicot and are considered to have more-ancient lineages.

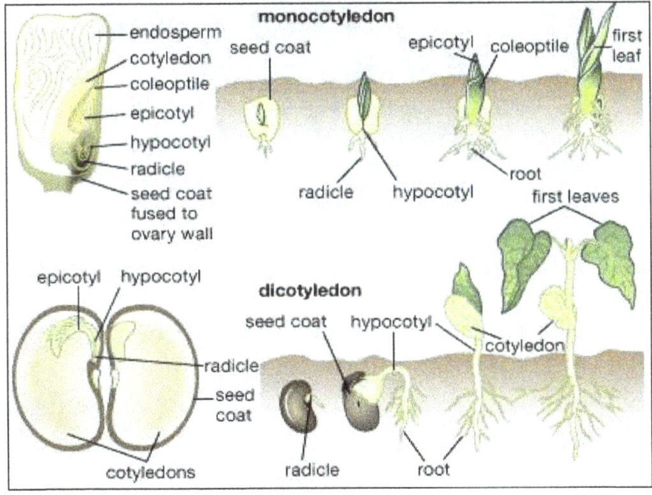

Germination of a monocot and a eudicot: (Top) In a corn seed (monocot), nutrients are stored in the cotyledon and endosperm tissue. The radicle and hypocotyl (region between the cotyledon and radicle) give rise to the roots. The epicotyl (region above the cotyledon) gives rise to the stem and leaves and is covered by a protective sheath (coleoptile). (Bottom) In a bean seed (eudicot), all nutrients are stored in the enlarged cotyledons. The radicle gives rise to the roots, the hypocotyl to the lower stem, and the epicotyl to the leaves and upper stem.

The plant body of angiosperms consists of a central axis of two parts, the shoot and the root. Shoots have two kinds of organs, the stem and the leaves, while roots have one type of organ, the root itself. Systems of classification are often based upon the longevity of the portions of plant aboveground. Woody plants are trees and shrubs whose shoots are durable and survive over a period of years. They are further classified into deciduous and evergreen plants. Deciduous plants drop their leaves at the end of every growing season, whereas evergreens keep their leaves for up to several years. Herbaceous plants have soft, flexible aerial portions and commonly die back each year.

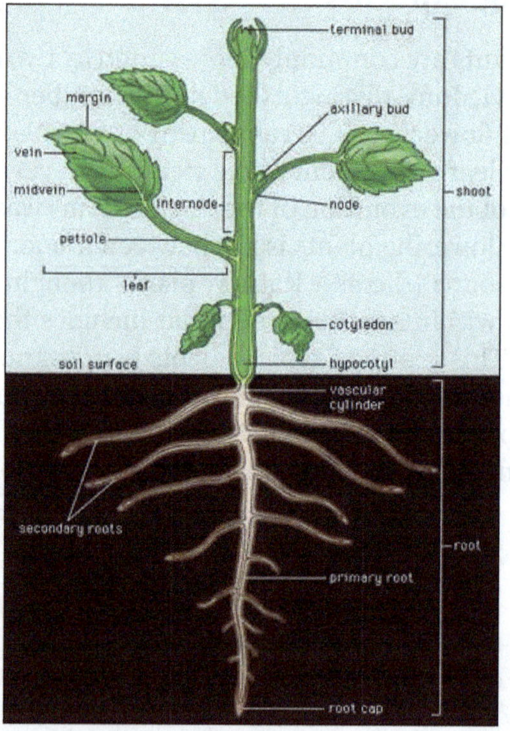

A typical dicotyledonous plant. (A dicotyledonous plant, or dicot, is any flowering plant that has a pair of leaves, or cotyledons, in the embryo of the seed.).

Another system of classification, based on the duration of the life history, is particularly applicable to angiosperms of the temperate region. Annuals are plants that complete the entire life history (germinate from seeds, mature, flower, and produce seed) in one growing season. Examples of annuals are corn, wheat, and peas. Biennials complete their life history in two seasons, blooming during the second season. Beets, celery, cabbage, carrots, and turnips are biennials, but their flowers are rarely seen because they are harvested during the first season. Annuals and biennials are both generally herbaceous plants. Perennials are plants that live from year to year. Trees and shrubs are perennials, but some herbaceous plants are also perennials.

Stems

Potato (*Solanum tuberosum*).

A number of modifications of the stem occur in angiosperms, and many of these modifications provide a means for herbs to become dormant and survive for a period of years. Rhizomes are horizontally growing underground stems that serve as organs of asexual reproduction and food storage. Similar to rhizomes, tubers are thickened underground stem portions that primarily serve as food storage (for example, potato). Corms are short upright underground stems surrounded by a few thin scale leaves (as in Crocus and Gladiolus). Bulbs have a greatly reduced stem with thick fleshy scale leaves surrounding it (as in the onion). Runners and stolons are surface stems characteristic of such plants as strawberries; new plants may form on the runner or stolon as it spreads along the ground. Many of the most prolific weeds have runners or stolons by which they propagate asexually.

Onion (Allium cepa).

In herbaceous dicotyledonous stems, the vascular conducting tissue (xylem and phloem) is organized into discrete strands or vascular bundles, each containing both xylem and phloem. The cells between the vascular bundles are thin-walled and often store starch. The peripheral region of cells in the stem is called the cortex; cells of the central portion make up the pith. The outermost cells of the stem compose the epidermis. No bark is formed on the herbaceous stem. In contrast, woody dicot stems develop an outer layer of dead thick-walled cells called cork cells, which together with the underlying phloem compose the bark of the tree. The major portion of the woody stem's diameter is a cylinder of xylem (wood) that originates from a region of cell division called the vascular cambium. The water-conducting cells that make up the xylem are nonliving. The accumulated xylem often forms annual rings composed of two zones: a relatively wide zone of spring wood (made up of large cells, characteristic of rapid growth) and a narrower zone of summer wood (smaller cells). Such rings may be absent in tropical trees that grow all year round. Xylem rays, radiating like spokes of a wagon wheel, are formed in the xylem and connect with the peripheral phloem. Stems of monocotyledons are composed of numerous vascular bundles that are arranged in a seemingly scattered manner within the ground tissue. Monocot vascular bundles lack a vascular cambium, and monocot stems thus do not become woody in a manner similar to dicots.

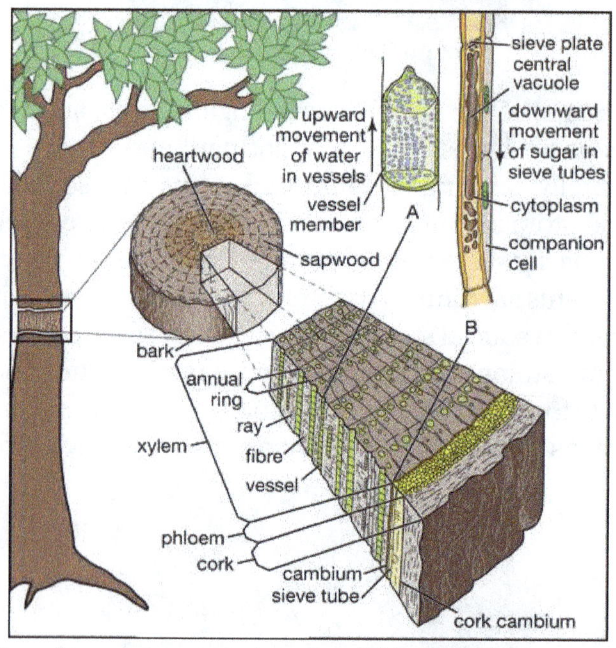

Internal transport system in a tree. (A) Enlarged xylem vessel. (B) Enlarged mature sieve element.

Leaves and Roots

Leaves are the other plant organ that, along with stems, constitutes the shoot of the vascular plant body. Their principal function is to act as the primary site of photosynthesis in the plant. Leaves of dicots possess a network of interconnecting veins and minor veins between the larger veins of the leaf (a pattern called net venation). Leaves of monocots possess major veins that extend parallel to the long axis of the leaf (parallel venation). Leaves are classified on the basis of leaf arrangement and whether they are simple or compound. A leaf may be deeply lobed but still simple; a compound leaf is composed of two or more distinctly separate leaflets.

Common leaf morphologies.

Structurally, leaves are composed of an outermost layer of cells called the epidermis. Epidermal cells secrete a waxy substance (cutin) that forms a cuticle impermeable to water. The pores (stomata) in the epidermis that allow for gas exchange are formed between specialized epidermal cells called guard cells. Vascular bundles (veins) are embedded in the mesophyll, the tissue that includes all of the cells between the upper and lower epidermis. The cells of the mesophyll contain the photosynthetic pigments.

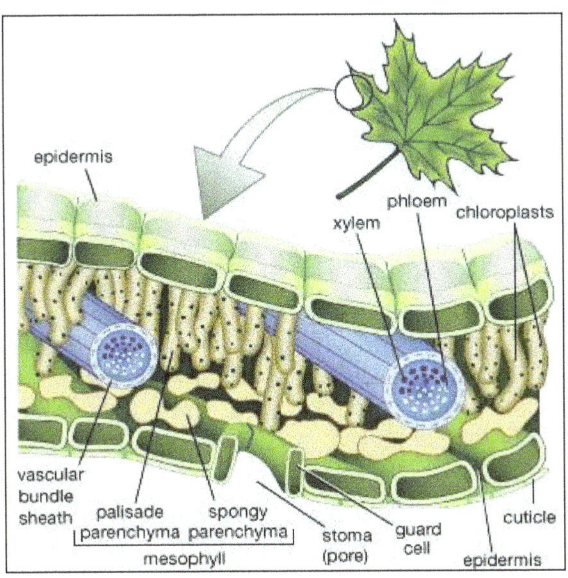

Structures of a leaf.

The epidermis is often covered with a waxy protective cuticle that helps prevent water loss from inside the leaf. Oxygen, carbon dioxide, and water enter and exit the leaf through pores (stomata) scattered mostly along the lower epidermis. The stomata are opened and closed by the contraction and expansion of surrounding guard cells. The vascular, or conducting, tissues are known as xylem and phloem; water and minerals travel up to the leaves from the roots through the xylem, and sugars made by photosynthesis are transported to other parts of the plant through the phloem. Photosynthesis occurs within the chloroplast-containing mesophyll layer.

The root system begins its development from the embryonic root (radicle), which grows out of the seed after the seed has absorbed water. This is the primary root of a new plant. The tip of the root is covered by a mass of loose cells called the root cap. Just beneath the root cap is the region of cell division of the root. Epidermal outgrowths just above the root tip are root hairs that are active in water and mineral absorption. Two types of root systems are commonly distinguished, fibrous roots and taproots. Fibrous root systems are composed of large numbers of roots nearly equal in size; root systems of this type are found, for example, in the grasses. A taproot system is one in which the primary root remains the largest, and a number of smaller secondary roots are formed from it; taproots are found in such plants as carrots and dandelions. Roots that arise other than by branching from the primary roots are called adventitious roots. The prop roots of corn, for example, are adventitious.

Two types of root system: (left) the fibrous roots of grass and (right) the fleshy taproot of a sugar beet.

Flowers

A primary distinction between the gymnosperms and the angiosperms is that the latter have flowers. Flowers represent modified shoots that have become differentiated for reproduction. The flower bears whorls of floral organs attached to a receptacle, the expanded end of a flower stalk on which the flower parts are borne. Sepals (collectively called the calyx) are modified leaves that encase the developing flower. They are sterile floral parts and may be either green or leaflike or composed of petal-like tissue. Petals (collectively called the corolla) are also sterile floral parts that usually function as visually conspicuous elements serving to attract specific pollinators to the flower. The calyx and the corolla together are referred to as the perianth. Flowers that lack one or both of the above perianth parts are called incomplete. Stamens (collectively called the androecium) are the male parts of the flower. Stamens are composed of saclike anthers (microsporangia) and filaments, which are stalks that support the anthers. Anthers are usually compartmentalized and contain the pollen grains (microgametophytes). The pistil, or female part of the flower, is composed of one or a number of carpels (collectively called the gynoecium) that fuse to form an essentially enclosed chamber. The three regions of the pistil (from the base up) are the ovary, which contains the ovules; the style, a stalked structure atop the ovary that elevates the stigma; and the stigma, a sticky knob whose surface receives the pollen during pollination.

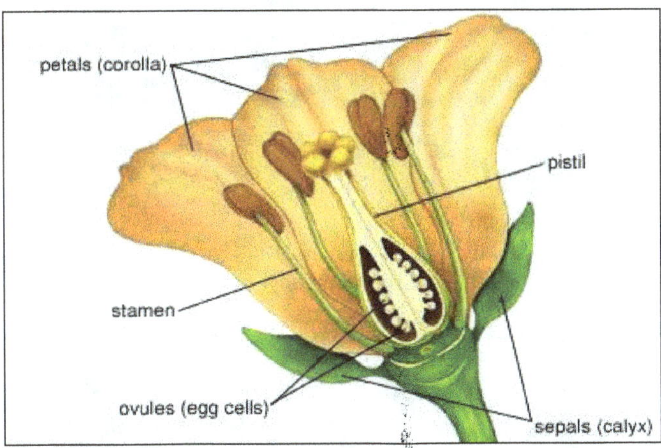

Diagram of a typical flowering plant (angiosperm).

Flowers may contain both male and female parts (a condition called perfect) or parts related to just one sex (imperfect), or they may have no sexual parts (sterile). Female and male flowers may be located on separate plants (dioecious) or on the same plant (monoecious). Flowers can also be borne singly or in aggregations called inflorescences.

Primitive flowers are radially symmetrical (actinomorphic) and are characterized by numerous spirally arranged floral parts. Floral parts are free (unfused) and are borne on an elongated floral axis. Sepals, petals, and stamens are attached below the ovary. Advanced flowers are bilaterally symmetrical and are characterized by a reduction in the number of floral parts. Floral parts are fused (often forming a long floral tube). Sepals, petals, and stamens are attached to the floral tube above the ovary.

Pollination is the transfer of pollen to the stigma of the same or another flower. Agents of pollination encompass a vast and diverse array of animals, including insects, birds, bats, honey possums, and slugs. Flowers exhibit various adaptations to pollinators, such as showy corollas, the production of nectar (a sugary liquid), and even visual cues visible only to insects that can perceive ultraviolet wavelengths of light. Flowers pollinated by wind generally are small and lack petals. The stigma is the pollen receptor site and must be chemically compatible with any pollen that lands on it for the pollen grain to germinate. This ensures that only genetically compatible sperm are transferred to the egg.

Fruits and Seeds

In flowering plants, ovules are enclosed and protected in an ovary. As the ovule develops into a seed, the ovary matures into a fruit. The formation of fruits is a characteristic feature of the flowering plants. Fruits are extremely variable. In some fruits, the ovary wall (pericarp) is thick and fleshy; in others, it is thin and dry.

Mature fruit of the papaya (Carica papaya).

Common wheat (Triticum aestivum).

Angiosperms have evolved many adaptations for seed dispersal involving such agents as wind, water, and animals. Adaptations to wind dispersal include wings or plumules attached to the seed or as part of the fruit or simply very minute seeds that are easily windborne. Adaptations to water dispersal are seeds that float or fruits that float and carry the seeds with them. Some seeds are a source of food to animals, which bury the seeds in the ground, where they later germinate. Other plants produce a fleshy fruit that is eaten along with the seeds inside it by animals, which pass the seeds through their digestive tracts unharmed. Another adaptation for animal dispersal is the development of barbed fruits or seeds that stick to the coats or skins of wandering animals. Some plants, such as witch hazels or jewelweed, can project their seeds through the air some distance from the parent plant.

Woolly seeds produced by the seed pods of the kapok tree (Ceiba pentandra).

Seedling of the red mangrove (Rhizophora mangle) ready to drop into the water after germinating on the tree.

Seeds have many adaptations that enable them to survive long periods of harsh conditions. Seeds can remain viable in a dormant condition for a few days or, in some species, for hundreds of years.

Reproduction and Life Histories

Each organism from inception to death goes through a sequence of genetically programmed developmental events constituting a life history. In eukaryotic organisms, development involves cellular events such as mitosis, meiosis, and syngamy (fertilization), which variously proceed by nuclear division (karyokinesis), cytoplasmic division (cytokinesis), cytoplasmic fusion without the union of nuclei (plasmogamy), or nuclear fusion (karyogamy).

Life Histories

The chromosome number in cells may be haploid, with one set of chromosomes per cell (written 1n); diploid, with two sets (2n); polyploid, with three or more sets; or dikaryotic, with a pair of nuclei in a cell (n + n), a condition that occurs mainly in fungi. Three types of sexual life histories have been recognized for the eukaryotic organisms: 1n, or haplontic; 2n, or diplontic; and 1n-2n (2n-1n). The former two types have collectively been called haplobiontic or monobiontic, because the life histories include only one phase; the third type has been called haplodiplontic, diplohaplontic, diplobiontic, dibiontic, or sporic, because the life history involves two alternating multicellular phases, or generations. Algae and fungi have many variants of all three types, especially the first, whereas land plants have the third type exclusively. In addition, all land plants are strictly oogamous, having motile sperm and nonmotile eggs. (In contrast, the algae and fungi may be oogamous or, frequently, isogamous or anisogamous, the latter conditions characterized by morphologically similar gametes that are either of the same size or with the female gametes of a larger size, respectively.).

The 1n-2n life history of bryophytes and vascular plants comprises the entire sequence of developmental events from zygote formation via syngamy (fertilization) to spore formation via meiosis. Syngamy and meiosis are successive events in a sexual life history. Syngamy involves the union of two 1n gametes to form a 2n zygote, which eventually develops into a 2n sporophyte. Meiosis involves the division of a 2n sporocyte (meiocyte, spore mother cell, pollen mother cell) to produce four 1n spores. These four spores constitute a tetrad. Gametes are 1n cells that fuse to form a zygote, whereas spores are 1n cells that develop into gametophytes without uniting with another cell.

Syngamy and meiosis are generally inseparable and alternate; consequently, the sporophyte develops from a zygote formed via syngamy, whereas the haploid gametophyte results after the sporophyte has undergone meiosis to form spores. These two phases, or generations, are multicellular in land plants. This type of life history involves an alternation of generations, a phenomenon not occurring in haplobiontic (1n or 2n) life histories. In the latter, such alternation is evident in both the morphological and the nuclear (chromosomal) changes that occur.

Meiosis and syngamy (fertilization) are the critical events that separate the sporophytic and gametophytic generations. (All nonmeiotic cell divisions involved in development are mitotic, in which chromosomes replicate, giving each daughter cell a full complement.) The zygote is the first stage and the sporocyte the last stage of the sporophytic generation, whereas the spore is the first stage and the gametes (eggs, sperm) the last stage of the gametophytic generation. The sporocytes of the multicellular 2n sporophyte divide meiotically to form 1n spores (sporogenesis). Each meiotic division results in a tetrad of four spores. (The land plants produce only sexual spores resulting from meiosis.) Each spore divides mitotically to form a multicellular 1n gametophyte, which eventually produces gametes mitotically (gametogenesis). The gametes (an egg and a sperm) fuse in the process of syngamy to form a 2n zygote. The zygote divides mitotically to form a multicellular embryo (embryogenesis), which is protected by either gametophytic tissues (such as remnants of archegonia in the nonseed land plants) or sporophytic tissues (the seed in the seed plants). An embryo, which is actually an immature sporophyte, is universally found among the land plants and often becomes dormant but eventually grows into the mature sporophyte. During organogenesis in vascular plants, the embryo develops into the mature sporophyte, with its vegetative organs (root and shoot, the latter consisting of stem and leaves) and reproductive structures (cones or strobili, flowers, etc.).

In all land plants, the two alternating generations are morphologically dissimilar. In all bryophytes, the gametophyte remains dominant, and the sporophyte is physiologically dependent on it. A homosporous life history, in which only one morphological type of spore is produced, is found in most bryophytes, although a few (e.g., Macromitrium) exhibit an anisosporous life history, in which the same sporangium produces morphologically similar spores of two different sizes.

Young sporophyte of tortula moss (Tortula muralis).

In vascular plants, the sporophyte ultimately becomes physiologically independent of the gametophyte. A homosporous life history occurs in Psilotum (Psilotophyta), Lycopodium (Lycophyta), Equisetum (Sphenophyta), and most ferns (Pteridophyta). A functionally heterosporous life history, in which the same sporangium produces morphologically similar but physiologically different spores, has been reported in a few pteridophytes—e.g., Equisetum. Finally, a heterosporous life history, in which different sporangia produce morphologically different types of spores, occurs in Selaginella, Isoetes (Lycophyta), a few aquatic ferns (Filicophyta), and all seed plants.

Homosporous Life Histories

A homosporous life history occurs in nearly all bryophytes and in most pteridophytes (lower vascular plants). It is characterized by morphologically identical spores that germinate to produce bisexual (both male and female) gametophytes in pteridophytes but either bisexual or, more usually, unisexual (either male or female) gametophytes in bryophytes. Each mature gametophyte bears gametangia (sex organs) that produce gametes. Each antheridium (male gametangium) forms many motile flagellate sperm, and each archegonium (female gametangium) forms one nonmotile egg. Fusion of an egg and a sperm (syngamy) creates a zygote and restores the 2n ploidy level. Various mechanisms prevent the fusion of eggs and sperm from a bisexual gametophyte (inbreeding). For example, the sex organs may mature at different times (usually antheridia mature first), or inbreeding may be chemically or genetically inhibited. The zygote divides mitotically to form the embryo, which then develops into the sporophyte. This eventually produces sporangia, which bear meiocytes (sporocytes) that divide meiotically to form spores. The number of spores produced per sporangium ranges from 16 or 32 in some pteridophytes to more than 65 million in some mosses. The sporangia may be borne in specialized structures, such as sori in ferns or as cones (strobili) in many other pteridophytes. The leaflike structures that bear sporangia are called sporophylls.

In most homosporous life histories of pteridophytes, the spores are both morphologically and physiologically identical and produce bisexual gametophytes. In some species of horsetail (Equisetum), the spores may be physiologically different and produce male or female gametophytes. This uncommon situation is called functional heterospory and may represent the means by which the heterosporous condition in vascular plants evolved from the homosporous condition.

Anisosporous Life Histories

In anisosporous life histories, an unusual phenomenon in bryophytes, there is a size difference between spores produced in the same sporangium. Each meiotic division results in a tetrad of two small spores that produce male gametophytes and two larger spores that produce female gametophytes.

Heterosporous Life Histories

A heterosporous life history occurs in some pteridophytes and in all seed plants. It is characterized by morphologically dissimilar spores produced from two types of sporangia: microspores, or male spores, and megaspores (macrospores), or female spores. In pteridophytes, megaspores are typically larger than microspores, but the opposite is true in most seed plants.

The spores produce two types of gametophytes: each microspore develops into a microgametophyte (male gametophyte), which ultimately produces male gametes (sperm), and each megaspore produces a megagametophyte (female gametophyte), which ultimately produces female gametes (eggs). Fusion of an egg and a sperm creates a zygote and restores the 2n ploidy level. The zygote divides mitotically to form the embryo, which then develops into the sporophyte. Eventually the sporophyte produces sporangia, which bear sporocytes (meiocytes) that undergo meiosis to form spores. Microsporangia (male sporangia) produce microsporocytes (micromeiocytes) that yield microspores. Megasporangia (female sporangia) produce megasporocytes (megameiocytes) that yield megaspores. The sporangia may be borne in specialized structures such as sori in ferns, cones (strobili) in some pteridophytes and most gymnosperms, or flowers in angiosperms. The leaflike structures bearing microsporangia and megasporangia are called, respectively, microsporophylls and megasporophylls. In angiosperms these sporophylls represent, respectively, the stamens and the carpels of the flower; in gymnosperms these sporophylls may constitute parts of, respectively, microstrobili (male cones, or pollen cones) and megastrobili (female cones, ovule cones, or seed cones).

Spruce tree (Picea species) laden with female cones.

The essential difference between the homosporous and heterosporous life history is the presence in the latter of two spore types (microspores and megaspores) and their concomitant precursory structures (microsporocytes and megasporocytes; microsporangia and megasporangia; etc.) and subsequent structures (microgametophytes and megagametophytes).

Variations involving Seed Plants

The gymnosperms and angiosperms not only lack some reproductive structures found in the homosporous and heterosporous pteridophytes but also have certain reproductive structures peculiar to the seed plants. Heterosporous pteridophytes, like their homosporous counterparts, have archegonia, antheridia, and motile flagellate sperm. The seed plants completely lack antheridia, and of the extant groups only the ginkgo and the cycads have flagellate sperm. Archegonia occur in most gymnosperms except Gnetum and Welwitschia, but they are lacking in all angiosperms.

Pollen grains and pollen tubes (male reproductive structures), ovules and seeds (female reproductive structures), and seedlings are structures unique to all seed plants. The ovule is a single megasporangium (in seed plants, this is called the nucellus) surrounded by one or two integuments (in rare cases, none or three) and containing inside the nucellus a single megasporocyte (spore mother cell). The megasporocyte undergoes meiosis to form four megaspores, three of which typically degenerate, the remaining one developing into the megagametophyte (female gametophyte). Ovules never dehisce (split open) to release their megaspores, unlike the megasporangia of most pteridophytes. The pollen grain is the partly or completely developed microgametophyte (male gametophyte). It is usually multicellular, consisting of two or three cells in angiosperms and usually two to five cells in gymnosperms, although in conifers it is occasionally one cell (for example, the families Taxaceae and some Cupressaceae) or 6 to 43 cells (the families Araucariaceae and some Podocarpaceae).

During pollination, pollen is transferred from its source to a receptive surface: in gymnosperms from the microsporangium to the integument or, especially, the pollination droplet of the ovule (rarely to the cone scale); in angiosperms from the microsporangium (pollen sac) of the anther to the stigma of the carpel. Once pollen has reached the appropriate receptive source, it germinates to form the pollen tube, a structure that grows toward the megagametophyte and in so doing conveys the sperm directly to the egg. All angiosperms and most gymnosperms, except ginkgo, cycads, and some fossil seed plants, lack swimming sperm. The presence of swimming sperm apparently represents a more primitive transitional evolutionary condition. After fertilization, the ovule transforms into a seed. The integument or integuments become modified into the seed coat. The seed typically becomes dormant for a period of time before it germinates to produce a seedling.

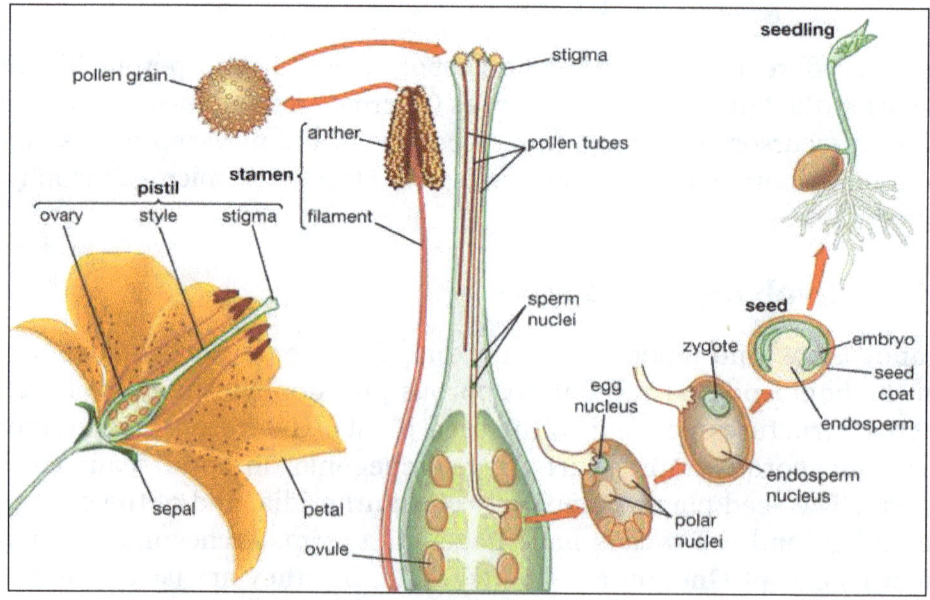

How flowering plants reproduce.

Reproduction in flowering plants begins with pollination, the transfer of pollen from anther to stigma on the same flower or to the stigma of another flower on the same plant (self-pollination) or from the anther on one plant to the stigma of another plant (cross-pollination). Once the pollen grain lodges on the stigma, a pollen tube grows from the pollen grain to an ovule. Two sperm nuclei then pass through the pollen tube. One of them unites with the egg nucleus and produces a zygote. The other sperm nucleus unites with two polar nuclei to produce an endosperm nucleus. The fertilized ovule develops into a seed.

Double fertilization is a phenomenon unique to angiosperms. Each pollen grain produces two sperm; one fuses with an egg to form the zygote, and the other fuses with one or more polar nuclei in the female gametophyte (megagametophyte, or also "embryo sac") to form an endosperm, which has a ploidy level that varies from 2n to 15n. In approximately 70 percent of the known cases, the second sperm fuses with two endosperm nuclei to produce a 3n (triploid) endosperm. The endosperm is a special nutritive tissue for the embryo and, after seed germination, for the seedling. In contrast, the megagametophyte is the comparable nutritive tissue in the gymnosperms.

Asexual Reproduction

Both homosporous and heterosporous life histories may exhibit various types of asexual reproduction (vegetative reproduction, somatic reproduction). Asexual reproduction is any reproductive process that does not involve meiosis or the union of nuclei, sex cells, or sex organs. Depending on the type of life history, asexual reproduction can involve the 1n or 2n generation.

The significance of sexual reproduction is that it is responsible for the genetic variation arising in a population as a result of the segregation and recombination of genetic material via meiosis and syngamy, respectively (the cells that result from sexual reproduction are genetically different from their parent cells). The significance of asexual reproduction is that it is a means for a rapid and significant increase in the numbers of individuals. (Many weeds and invasive species, for instance, are successful partly because of their great capacity for vegetative reproduction.) The cells that result from asexual reproduction are genetically identical to their parent cells. In addition, vegetative reproduction in the bryophytes and pteridophytes is a means of bypassing the somewhat lengthy and moisture-dependent sexual process; that is, the motile swimming sperm characteristic of these groups require the presence of water, which may be a limiting factor in drier times.

Deviations from the Usual Life History

In most life histories, a 2n sporophyte typically alternates with a 1n gametophyte, but there are significant deviations. Apospory is the development of 2n gametophytes, without meiosis and spores, from vegetative, or nonreproductive, cells of the sporophyte. In contrast, apogamy is the development of 1n sporophytes without gametes and syngamy from vegetative cells of the gametophyte. The 2n aposporous gametophytes and the 1n apogamous sporophytes are usually infertile under natural conditions because of disruption of cytological events. Various compensating genetic mechanisms, however, may occur to complete the life history. Parthenogenesis is the formation of a 1n embryo directly from an unfertilized egg. Apospory and apogamy occur in bryophytes, pteridophytes, and angiosperms, whereas parthenogenesis occurs in ferns and angiosperms. Apogamy is more common in pteridophytes, but apospory is more common in bryophytes.

Some ferns (certain species of Trichomanes and Vittaria) have lost the ability to produce sporophytes. The species exist as gametophytes that spread by gemmae (units of asexual reproduction); although gametangia are produced, no sporophytes result.

Plant Physiology

General Features of Plant Nutrition

Plant nutrition includes the nutrients necessary for the growth, maintenance, and reproduction of individual plants; the mechanisms by which plants acquire such nutrients; and the structural, physiological, and biochemical roles those nutrients play in metabolism.

Mode of Nutrition

All organisms obtain their nutrients from the environment, but not all organisms require the same nutrients, nor do they assimilate these nutrients in the same way. There are two basic nutritional types, autotrophs and heterotrophs. Heterotrophs require both inorganic and organic (carbon-containing) compounds as nutrient sources. Autotrophs obtain their nutrients from inorganic compounds, and their source of carbon is carbon dioxide (CO_2). An autotroph is photoautotrophic if light energy is required to assimilate CO_2 into the organic constituents of the cell. Furthermore, a photoautotroph that also uses water and liberates oxygen in the energy-trapping process of photosynthesis is an oxygenic photoautotroph. Earth's first such organisms are believed to have been the major sources of the present-day oxygen content of the atmosphere (approximately 21 percent). Almost all plants, as well as many prokaryotes and protists, are characteristically oxygenic photoautotrophs.

Plants, as autotrophic organisms, use light energy to photosynthesize sugars from CO2 and water. They also synthesize amino acids and vitamins from carbon fixed in photosynthesis and from inorganic elements garnered from the environment. (Animals, as heterotrophic organisms, cannot synthesize many nutrients, including certain amino acids and vitamins, and so must take them from the environment.).

Essential Elements and Minerals

Certain key elements are required, or essential, for the complex processes of metabolism to take place in plants. Plant physiologists generally consider an element to be essential if (1) the plant is unable to complete its life cycle (i.e., grow and reproduce) in its absence; (2) the particular structural, physiological, or biochemical roles of the element cannot be satisfied by any other element; and (3) the element is directly involved in the plant's metabolism (e.g., as part of an enzyme or other essential organic cellular constituent). Beneficial elements are those that stimulate plant growth by ameliorating the toxic effects of other elements or by substituting for an element in a less-essential role (e.g., as a nonspecific osmotic solute). Some elements are beneficial in that they are necessary for the growth of some, but not all, plant species.

The required concentrations of each essential and beneficial element vary over a wide range. The essential elements required in relatively large quantities for adequate growth are called macroelements. Nine minerals make up this group: carbon (C), hydrogen (H), oxygen (O), nitrogen (N), potassium (K), calcium (Ca), magnesium (Mg), phosphorus (P), and sulfur (S). Eight other essential mineral elements are required in smaller amounts (0.01 percent or less) and are called microelements. These are iron (Fe), chlorine (Cl), manganese (Mn), boron (B), copper (Cu), molybdenum (Mo), zinc (Zn), and nickel (Ni). The specific required percentages may vary considerably with species, genotype (or variety), age of the plant, and environmental conditions of growth.

A macronutrient is the actual chemical form or compound in which the macroelement enters the root system of a plant. The macronutrient source of the macroelement nitrogen, for example, is the nitrate ion (NO_3^-); alternatively, nitrogen is taken up as the ammonium ion (NH_4^+) or as amino acids. Carnivorous plants use nitrogen from the proteins and nucleic acids of the prey they catch. Carbon dioxide from the atmosphere provides the carbon and oxygen atoms. Water taken from the soil provides much of the hydrogen. Soil provides macroelements and microelements from mineral complexes, parent rock, and decaying organisms. Factors that determine plant root uptake include the solubility and mobility of the chemical in question, the adsorptive properties of the charged soil surfaces, and the surface area and uptake capacity of the roots of the individual plant.

The macroelements carbon, hydrogen, oxygen, and nitrogen constitute more than 96 percent of the dry weight of plants. Thus, they are the major constituents of the structural and metabolic compounds of the plant. Their presence and that of potassium within cells also helps regulate osmotic pressure. In addition, phosphate is a constituent of nucleic acids, including DNA, and membranes; it also plays a role in various metabolic pathways. Microelements are generally either activators or components of enzymes, although the macroelements potassium, calcium, and magnesium also serve these roles.

Metabolic Cycles

Metabolism denotes the sum of the chemical reactions in the cell that provide the energy and synthesized materials required for growth, reproduction, and maintenance of structure and function. In plants the ultimate source of all organic chemicals and the energy stored in their chemical bonds is the conversion of CO_2 into organic compounds (CO_2 fixation) by either photosynthesis or chemosynthesis. The general and specific features of plant metabolism ultimately derive from oxygenic photosynthesis, which underlies the autotrophic nutrition of plants.

Pathways and Cycles

Chemical reactions in the cell occur in a sequence of stages called a metabolic pathway. Each stage is catalyzed by an enzyme, a protein that changes (usually increases) the rate at which the reaction proceeds but does not alter the reactants or end products. Certain thermodynamic conditions must be met for a reaction to proceed, even in the presence of enzymes. If the end product of the reaction is also the reactant (or substrate) that starts the pathway, then the sequence of reactions is called a metabolic cycle. The intermediate chemicals that are formed and used in the various stages of the sequence are called intermediary metabolites.

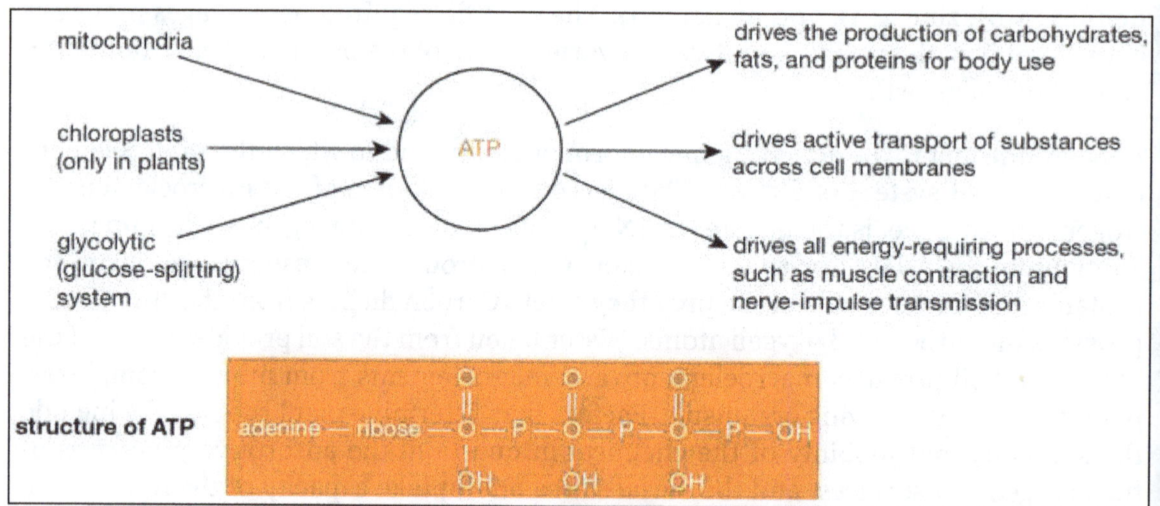

Adenosine Triphosphate; Physiology.

Metabolic pathways and cycles are either catabolic (energy-releasing) or anabolic (energy-consuming). Catabolic reactions break down complex metabolites into simpler ones, whereas anabolic reactions build up (biosynthesize) new molecules. When chemical bonds are broken, energy is released, which drives anabolic reactions to form new bonds. The energy released generally has been stored in high-energy bonds of an intermediate energy carrier molecule, such as the terminal phosphate bond of adenosine triphosphate (ATP). (When the terminal phosphate is split from the ATP molecule, adenosine diphosphate, or ADP, is formed and inorganic phosphate is released, along with energy.). The simpler metabolites formed via catabolic reactions are often the building-block metabolites used in anabolic reactions to synthesize more complex molecules (e.g., starch, proteins, or lipids.).

Adenosine triphosphate (ATP) is the power source of many biochemical reactions. It is produced in the cell structures and system listed at the left to energize the important life processes listed on the right. An abbreviated chemical formula of the structure of ATP is also shown. The two high-energy $P-O-P$ bonds are responsible for its power.

Control Mechanisms

The cells of all plants are eukaryotic, because they possess a nucleus and membrane-bound organelles, such as chloroplasts, mitochondria, glyoxysomes, peroxisomes, and vacuoles. The thousands of metabolic reactions that take place in the cell are regulated within these organelles and their subcompartments. When compared with cells of other eukaryotic organisms, plant cells have a high degree of metabolic compartmentalization.

The primary mechanism of metabolic control, however, remains the enzymes themselves. Although all enzymes of the pathway help determine the net and directional flow of carbon, certain key stages are controlled by regulatory enzymes. Regulatory enzymes may either catalyze the first stage in the metabolic pathway or catalyze reactions in which key branch points occur. The activity of such enzymes, in turn, may be controlled by the amount synthesized (coarse control by gene expression)—in that further action by the enzyme is inhibited when some critical concentration of the reaction product is reached—or by special metabolites, called effectors, that interact directly with the enzyme (fine control). The latter metabolites may be either part of or totally unrelated to the metabolites of the pathway.

Another mechanism by which metabolic reactions are regulated is through transport systems in the membranes of organelles. These systems control the nature, direction, and amount of metabolites entering the metabolic pathways and are often uniquely related to the autotrophic nutrition of plants.

Principal Pathways and Cycles

The 6-carbon sugar glucose, a product of photosynthesis, is mostly translocated in the form of sucrose (a 12-carbon sugar) to nourish nonphotosynthesizing parts of the plant, or it may be polymerized into starch for storage. (Trehalose, another 12-carbon sugar, replaces sucrose in some vascular plants; others transport even larger sugars or sugar alcohols.) When required, sucrose and starch are hydrolyzed to glucose and then enter glycolysis or the pentose phosphate pathway. The reactions of both pathways take place in the cytoplasm of the cell.

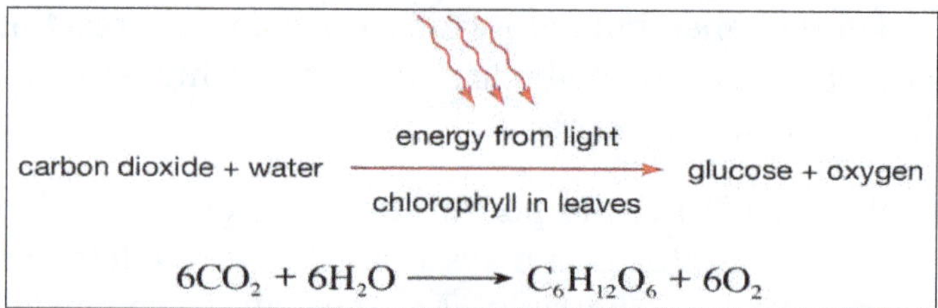

$$6CO_2 + 6H_2O \longrightarrow C_6H_{12}O_6 + 6O_2$$

In photosynthesis, plants consume carbon dioxide and water and produce glucose and oxygen. Energy for this process is provided by light, which is absorbed by pigments, primarily chlorophyll. Chlorophyll is the pigment that gives plants their green colour.

The net result of glycolysis is the metabolism of glucose into two molecules of the four-carbon organic acid malate. This metabolic pathway involves phosphate-containing intermediates and is regulated by two enzymes, which catalyze those reactions that contain the substrates fructose phosphate and phosphoenolpyruvate (PEP). Glycolysis yields ATP molecules and hydrogen; the latter is accepted by the coenzyme (coenzymes are smaller, nonprotein participants associated with certain enzymes) nicotinamide adenine dinucleotide (NAD) to form NADH. The hydrogen on NADH then reacts either with molecular oxygen (O_2) to capture the energy (and transfer it to the high-energy bonds of ATP) or with another metabolite to reduce the molecule by the addition of hydrogen. Some intermediates are used in the biosyntheses of fat or certain amino acids.

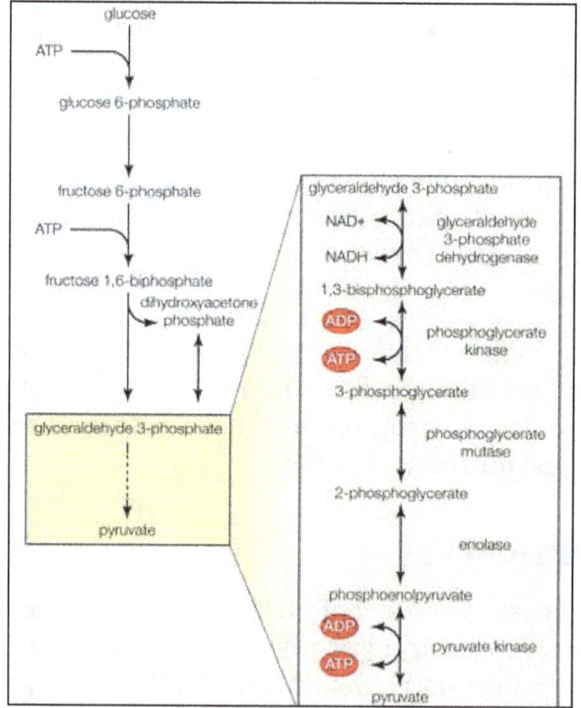

The generation of pyruvate through the process of glycolysis is the first step in fermentation.

The pentose phosphate pathway is an alternative pathway for the catabolism of glucose, producing end products that are used in the biosynthesis of nucleic acids, some vitamins, and key metabolites. It also furnishes reducing power (i.e., it accepts hydrogen atoms and carries them on the coenzyme nicotinamide adenine dinucleotide phosphate [NADP]) for use in the synthesis of substances such as fat. It is regulated by the rate at which the product of the pentose pathway, NADPH, is oxidized.

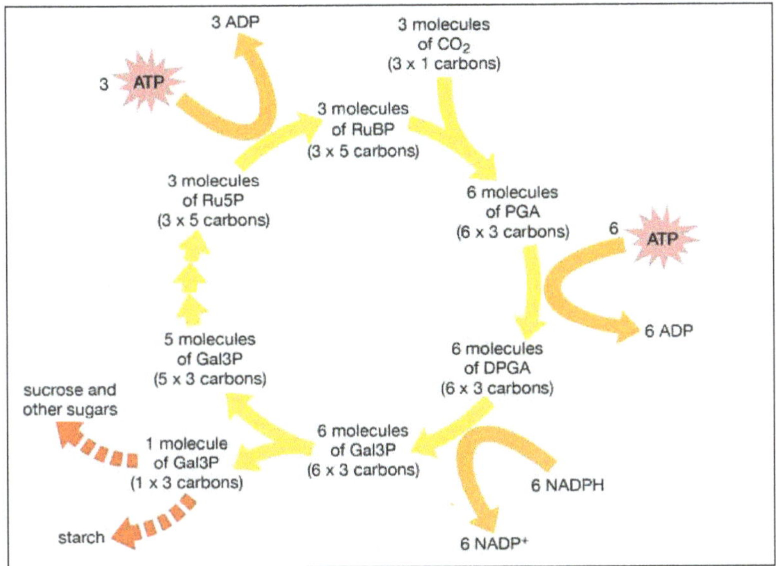

C$_3$ carbon fixation pathway.

Pathway of carbon dioxide fixation and reduction in photosynthesis, the reductive pentose phosphate cycle. The diagram represents one complete turn of the cycle, with the net production of one molecule of Gal3P. The nine molecules of ATP and six molecules of NADPH come from the light reactions.

Malate produced in glycolysis is transported into the mitochondria, where it enters a sequence of 10 reactions called the tricarboxylic acid (TCA) cycle, or Krebs cycle. Malate is converted into pyruvate, which is then metabolized into the two-carbon intermediate, acetyl coenzyme A (CoA), which combines with a four-carbon acid, oxaloacetate. The product, citrate, has three carboxylic acid groups—hence the name tricarboxylic acid cycle. Citrate is systematically catabolized (broken down) with progressive losses of successive carbon atoms as CO$_2$ into five-carbon and, finally, four-carbon, acids. The latter acid, oxaloacetate, begins the cycle again. With each oxidation reaction, a hydrogen atom is transferred to the coenzyme NAD or, in one reaction, the coenzyme flavin adenine dinucleotide (FAD) to form NADH and FADH, respectively. The reduced coenzymes NADH and FADH enter into a sequence of reactions called the respiratory chain on the inner membrane of the mitochondrion. This chain is a series of carriers (ubiquinone and several iron-containing chemicals called cytochromes) that ultimately transfer the hydrogen and electrons of these coenzymes to molecular oxygen, forming water. The energy generated from the oxidation by the respiratory chain is trapped in three ATP molecules formed per NADH molecule oxidized. The mechanism is chemiosmotic in that it involves building a hydrogen ion (proton) gradient on one side of the mitochondrial membrane.

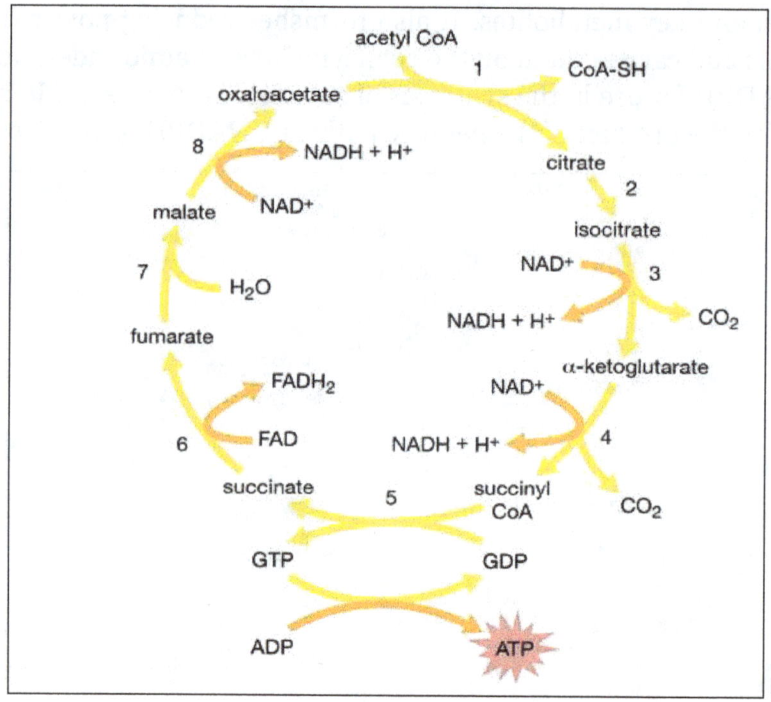

The eight-step tricarboxylic acid cycle.

A net of 36 ATP molecules are gained from all hydrogen-carrying coenzymes formed in glycolysis and the TCA cycle, and they represent the principal energy source for most anabolic (biosynthetic) reactions in plants. In addition, the TCA cycle furnishes metabolites for the biosynthesis of important organic molecules of the cell.

Another metabolic cycle, the isoprenoid pathway, produces essential oils, carotenoid pigments, certain plant hormones, and rubber. These metabolites are unique to plants and serve such functions as attracting pollinating insects, providing defense against herbivores, and producing photosynthetic pigments and phytohormones. Plant seedlings use the glyoxylic acid cycle to convert fats (principally from seeds) into glucose. This occurs initially in the glyoxysome and subsequently in the mitochondria and cytosol (the fluid mass that surrounds the various organelles).

Unique Features of Plant Metabolism

The pathways outlined above exist in essentially the same form in all organisms, but metabolism in plants does have certain unique features. Plant mitochondria, for example, have specific transport systems for the NADH produced in glycolysis and for the oxaloacetate produced from a direct fixation of CO_2 into PEP. Unlike animal mitochondria, plant mitochondria metabolize malate and the amino acid glycine. A special enzyme converts malate to pyruvate, thereby allowing an alternative to the glycolytic pathway that is common in other organisms. Glycine is a product of the unique plant pathway of photorespiration.

Plant mitochondria possess a cyanide-resistant alternative respiratory chain in addition to the cyanide-sensitive cytochrome chain also found in other organisms. Oxidation of NADH through this alternative pathway produces energy in the form of heat but no ATP. Some physiologists suggest that this pathway is a mechanism to prevent overreduction of the respiratory pathway, which would lead to the production of toxic free radicals. Others believe that this pathway allows the TCA cycle to continue at times of decreased need for ATP, to produce more than the usual amount of metabolites, which, in the presence of ATP, could not normally be produced. This system functions at a high rate in the flowers of a range of species, including the arum lily (Araceae). Temperatures of this organ may reach 40 °C (104 °F), which also contributes to the attraction of pollinators.

Photosynthesis

The autotrophic mode of nutrition of plants, is derived from oxygenic photosynthesis. Energy-rich organic compounds are synthesized from low-energy atmospheric CO_2, using the energy of absorbed sunlight. (Some bacteria are nonoxygenic photosynthesizers, utilizing hydrogen sulfide, H_2S, rather than water.) The resultant organic compounds initiate the flow of energy and carbon through the food chains of managed and natural ecosystems, intrinsically linking plants with the heterotrophic life-forms of the remaining kingdoms of organisms. The oxygen liberated by plants (and certain photosynthetic protists and prokaryotes) over geologic time has oxygenated Earth's atmosphere and has produced fossil fuels such as coal, gas, and oil.

The following sections describe the basic mechanisms of photosynthesis—the acquisition of energy and the fixation of carbon dioxide—used by plants of diverse evolutionary lines.

Basic Mechanisms

Electromagnetic radiation having wavelengths between approximately 400 and 700 nanometres can be seen as light by the eye and constitutes the range absorbed by plants for photosynthesis. Blue light has a wavelength around 450 nanometres, and red light, a wavelength of 650–700 nanometres.

Double-membraned cell organelles called chloroplasts contain the photosynthetic apparatus: light-absorbing pigments, other electron-carrying chemicals (cytochromes and quinones), and enzymes. (Pigments absorb light of a particular wavelength; those wavelengths that are not absorbed are reflected and may be perceived as colour—hence, for example, the green colour of many plants.). The inner membrane of the chloroplast is folded into flat tubes, the edges of which are joined to hollow sack-like disks called thylakoids. Stacks of thylakoids embedded with pigment molecules are called grana. The inner matrix of the chloroplast is called the stroma.

The internal (thylakoid) membrane vesicles are organized into stacks, which reside in a matrix known as the stroma. All the chlorophyll in the chloroplast is contained in the membranes of the thylakoid vesicles.

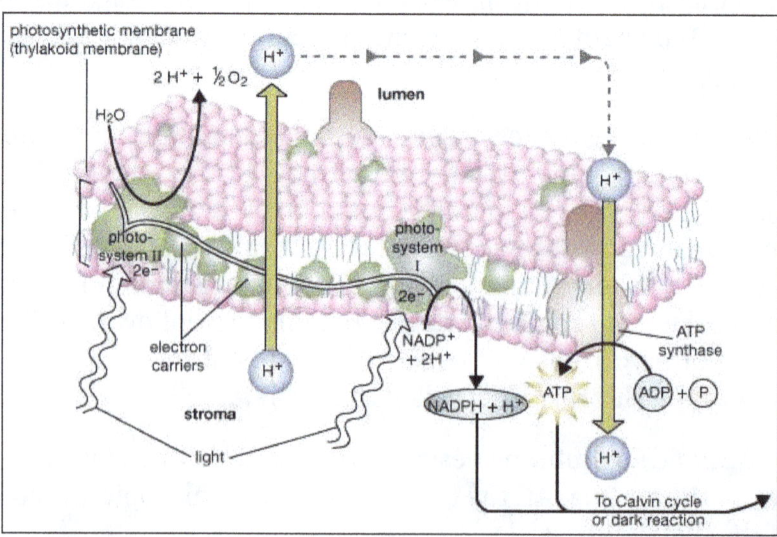

Chloroplast structure.

Photosynthesis consists of two interdependent series of reactions, the light, or light-harvesting, reactions and the dark, or carbon-assimilating, reactions; the former are dependent on light, the latter on temperature. Light reactions occur in the grana and dark reactions in the stroma. The overall formula for photosynthesis is:

$$6CO_2 + 12H_2O \rightarrow C_6H_{12}O_6 + 6O_2 + 6H_2O$$

The light reactions, the first stage of photosynthesis, convert light energy into chemical energy (ATP and NADPH). Light reactions comprise two interdependent systems, called photosystems I and II. The dark reactions, the second stage of photosynthesis, use the chemical energy products of the light reactions to convert carbon from carbon dioxide to simple sugars.

The light reaction of photosynthesis: The light reaction occurs in two photosystems (units of chlorophyll molecules). Light energy (indicated by wavy arrows) absorbed by photosystem II causes the formation of high-energy electrons, which are transferred along a series of acceptor molecules in an electron transport chain to photosystem I. Photosystem II obtains replacement electrons from water molecules, resulting in their split into hydrogen ions (H^+) and oxygen atoms. The oxygen atoms combine to form molecular oxygen (O_2), which is released into the atmosphere. The hydrogen ions are released into the lumen. Additional hydrogen ions are pumped into the lumen by electron acceptor molecules. This creates a high concentration of ions inside the lumen. The flow of hydrogen ions back across the photosynthetic membrane provides the energy needed to drive the synthesis of the energy-rich molecule ATP. High-energy electrons, which are released as photosystem I absorbs light energy, are used to drive the synthesis of NADPH. Photosystem I obtains replacement electrons from the electron transport chain. ATP provides the energy and NADPH provides the hydrogen atoms needed to drive the subsequent photosynthetic dark reaction, or Calvin cycle.

Light reactions consist of several hundred light-absorbing pigment molecules arranged so as to maximize the gathering of light energy. These "antennae" are coupled to a mini-circuit of electron-carrying chemicals. The pigments are two types of chlorophyll—chlorophyll a and chlorophyll b—and various carotenoids. Absorbed light energy is transferred to specialized chlorophyll molecules called P_{700} and P_{680} in photosystems I and II, respectively. Once these specialized chlorophyll molecules have acquired sufficient energy, electrons are given up to the electron carriers within their photosystems, initiating an electron flow. (The carrier molecules include plastoquinones and cytochromes.). The effect of this, when photosystems I and II function synchronously, is the formation of a chemiosmotic gradient of protons that phosphorylates (adds a phosphate group to) ADP, resulting in ATP. Those electrons also lead to the formation of NADPH from NADP. The P_{680} chlorophyll, upon loss of its electron, becomes a strong oxidizing agent that subsequently causes the water molecule to dissociate into protons and oxygen gas.

The dark reactions are responsible for the conversion of carbon dioxide to glucose. The essential reaction involves the combining of CO_2 with the five-carbon sugar ribulose 1,5-bisphosphate (RuBP) in a series of reactions called the Calvin-Benson cycle. This reaction yields an unstable six-carbon intermediate, which immediately breaks down into two molecules of phosphoglycerate (PGA), a three-carbon acid. Each reaction is catalyzed by a specific enzyme. Six revolutions of the cycle means that 6 CO_2 molecules react with 6 RuBP molecules to produce 12 molecules of PGA; 2 three-carbon PGA molecules combine to form the six-carbon glucose, and 10 PGAs are recycled to regenerate 6 molecules of RuBP. The ATP and NADPH from light reactions provide the energy and reducing power to form glucose and refurbish the CO_2 acceptor, RuBP.

Specific Variations in Photosynthesis

Chlorophylls a and b (bound to proteins) and carotenoids constitute the principal light-absorbing complex of most plants. Differences in chloroplast structure, though not major, occur among phylogenetically diverse plant groups. All such variations, however, represent evolutionary adaptations to more efficiently utilize the light energy that drives the reactions common to all oxygenic photosynthesizers (i.e., photosystems I and II) or to avoid damage due to excessive light.

The enzyme ribulose 1,5-bisphosphate carboxylase/oxygenase (Rubisco) catalyzes the formation of organic molecules from CO_2. As the major enzyme of all photosynthetic cells, Rubisco is the most abundant protein on Earth. There is, however, a major catalytic flaw in the ability of this enzyme to convert CO_2 to sugars. In the presence of molecular oxygen, Rubisco also catalyzes a reaction in which oxygen is introduced (i.e., it acts as an oxygenase), and CO_2 is formed rather than converted.

Rubisco evolved in photosynthetic organisms that lived in the atmosphere of primitive Earth, an atmosphere that contained only traces of molecular oxygen and plenty of carbon dioxide. As photosynthesis in the cyanobacteria of Precambrian times (about 4.6 billion years ago to 541 million years ago) oxygenated the atmosphere, the ratio of carbon dioxide to oxygen fell drastically, and Rubisco began to function more and more as an oxygenase. This greatly reduced the net fixation of CO_2 into sugars and, therefore, photosynthetic efficiency. Rubisco as an oxygenase splits RuBP into one PGA and a two-carbon acid phosphoglycolate, which initiates the photorespiratory carbon-oxidation cycle, or photorespiration. (This cycle probably evolved to recycle PGA back into the photosynthetic pathway, thereby preventing an even greater loss of carbon.) Photorespiration involves three organelles (chloroplasts, peroxisomes, and mitochondria), each with unique transport mechanisms for the cycle's intermediates.

All plants are classified as C_3 (plants that use only the Calvin-Benson cycle), C_4 (plants that use an additional CO_2-fixation mechanism and the Calvin-Benson cycle), C_3-C_4 (plants intermediate between C_3 and C_4), and CAM (plants that have a nocturnal variant of the C_4 pathway).

The majority of plants fix CO_2 directly into RuBP, and their first stable product is the three-carbon acid PGA—hence the designation C_3. Those plants have an active photorespiratory cycle, especially at high temperatures.

Sometime during the Oligocene Epoch (33.9 million to 23 million years ago), certain of the angiosperms (grasses and dicotyledonous plants) of mainly tropical climates evolved a CO_2-fixation system that acted ahead of the Calvin-Benson cycle. The first fixation is into the three-carbon acid phosphoenolpyruvate (PEP) by PEP carboxylase (an enzyme that has no oxygenase function) in the outer mesophyll cells of the leaf. The first stable fixation product is the four-carbon acid oxaloacetate—hence the designation C_4 plants. Oxaloacetate is reduced to malate, which is transferred to a thick-walled bundle sheath cell. Malate is decarboxylated, giving rise to high CO_2 concentrations in the bundle sheath. Here, Rubisco of the Calvin-Benson cycle functions more efficiently because oxygenation is suppressed. There is thus a spatial separation of initial CO_2 fixation and the Calvin-Benson cycle. This efficiency is not without cost, however, as additional ATP is required to recycle PEP. For this reason, C_3 plants may be more efficient in cold climates, where photorespiration is insignificant; under conditions where there is less available light, the higher ATP requirement would become a penalty.

The C_4 pathway is effective at fixing CO_2 under drought conditions or under conditions where CO_2 is limited. C_4 plants do not need to open their stomata as wide as C_3 plants, because their primary carboxylating enzyme is saturated at much lower CO_2 concentrations. As a result, they lose less water during photosynthesis, and they are better able to cope in regions with arid climates. As humans continue to burn fossil fuels and thus increase the CO_2 concentration in the atmosphere, the relative advantage C_4 plants enjoy in Earth's warm regions at present will diminish.

There are also plants with enzymatic and leaf anatomical characteristics intermediate between C_3 and C_4 plants, called C_3-C_4 intermediate species. Those plants are thought to be in the pathway of evolution to full C_4 photosynthetic status.

Bishop's cap cactus (*Astrophytum myriostigma*).

Succulent plants of the desert regions (e.g., cacti) also initially fix CO_2 into oxaloacetate. This occurs only at night when conditions are cooler, however. Normally, the stomata in leaves or stems, through which plants lose water and acquire carbon dioxide, are open in the day and closed at night; however, the stomates of succulent plants that use the C_4 pathway do the opposite and hence prevent loss of water during the hot days. The resultant oxaloacetate is converted into malate, stored in the vacuole as malic acid, and released during the day when the stomates are closed. Malate is decarboxylated, and the CO_2 that is released is fixed by Rubisco in the usual Calvin-Benson cycle. Both the C_4 and C_3 processes take place in the same cell. This process is called crassulacean acid metabolism (hence CAM plants), after a family of succulent plants (Crassulaceae).

Plant Anatomy

Plant anatomy is the study of the shape, structure, and size of plants. As a part of botany (the study of plants), plant anatomy focuses on the structural or body parts and systems that make up a plant. A typical plant body consists of three major vegetative organs: the root, the stem, and the leaf, as well as a set of reproductive parts that include flowers, fruits, and seeds.

As a living thing, all of a plant's parts are made up of cells. Although plant cells have a flexible membrane like animal cells, a plant cell also has a strong wall made of cellulose that gives it a rigid shape. Unlike animal cells, plant cells also have chloroplasts that capture the Sun's light energy and convert it into food for itself. Like any complex living thing, a plant organizes a group of specialized cells into what are called tissues that perform a specific function. For example, plants therefore have epidermal tissue that forms a protective layer on its surface. They also have parenchyma tissue usually used to store energy. The "veins" or pipeline of a plant are made up of vascular tissue that distribute water, minerals, and nutrients throughout the plant. Combined tissues form organs that play an even more complex role.

Plant Cells and Tissues 2

- Plant Cell
- Nuclear Envelope
- Cell Wall
- Central Vacuole
- Cell Nucleus
- Mitochondria
- Plastid
- Golgi Apparatus
- Endoplasmic Reticulum
- Phragmoplast
- Plant Tissue
- Meristem
- Vascular Tissue
- Epidermis
- Ground Tissue

The eukaryotic cells present in green plants that belong to the kingdom Plantae are referred to as plant cells. Plants have three tissue types – dermal, ground and vascular tissues. This chapter has been carefully written to provide an easy understanding of all the aspects related to plant cells and tissues.

Plant Cell

Plant cells are eukaryotic cells or cells with a membrane-bound nucleus. Unlike prokaryotic cells, the DNA in a plant cell is housed within a nucleus that is enveloped by a membrane. In addition to having a nucleus, plant cells also contain other membrane-bound organelles (tiny cellular structures) that carry out specific functions necessary for normal cellular operation. Organelles have a wide range of responsibilities that include everything from producing hormones and enzymes to providing energy for a plant cell.

Plant cells are similar to animal cells in that they are both eukaryotic cells and have similar organelles. However, there are a number of differences between plant and animal cells. Plant cells are generally larger than animal cells. While animal cells come in various sizes and tend to have irregular shapes, plant cells are more similar in size and are typically rectangular or cube shaped. A plant cell also contains structures not found in an animal cell. Some of these include a cell wall, a large vacuole, and plastids. Plastids, such as chloroplasts, assist in storing and harvesting needed substances for the plant. Animal cells also contain structures such as centrioles, lysosomes, and cilia and flagella that are not typically found in plant cells.

Plant Cell Organelles

The following are examples of structures and organelles that can be found in typical plant cells:

- Cell (Plasma) Membrane: This thin, semi-permeable membrane surrounds the cytoplasm of a cell, enclosing its contents.

- Cell Wall: This rigid outer covering of the cell protects the plant cell and gives it shape.

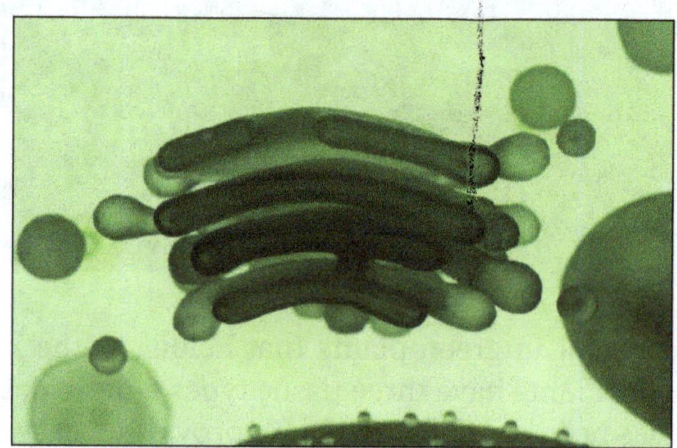

The Golgi Apparatus Model.

- Chloroplast: Chloroplasts are the sites of photosynthesis in a plant cell. They contain chlorophyll, a green pigment that absorbs energy from sunlight.

- Cytoplasm: The gel-like substance within the cell membrane is known as cytoplasm. It contains water, enzymes, salts, organelles, and various organic molecules.

- Cytoskeleton: This network of fibers throughout the cytoplasm helps the cell maintain its shape and gives support to the cell.

- Endoplasmic Reticulum (ER): The ER is an extensive network of membranes composed of both regions with ribosomes (rough ER) and regions without ribosomes (smooth ER). The ER synthesizes proteins and lipids.

- Golgi Complex: This organelle is responsible for manufacturing, storing and shipping certain cellular products including proteins.

- Microtubules: These hollow rods function primarily to help support and shape the cell. They are important for chromosome movement in mitosis and meiosis, as well as cytosol movement within a cell.

- Mitochondria: Mitochondria generate energy for the cell by converting glucose (produced by photosynthesis) and oxygen to ATP. This process is known as respiration.

- Nucleus: The nucleus is a membrane-bound structure that contains the cell's hereditary information (DNA).

  - Nucleolus: This structure within the nucleus helps in the synthesis of ribosomes.

  - Nucleopore: These tiny holes within the nuclear membrane allow nucleic acids and proteins to move into and out of the nucleus.

- Peroxisomes: Peroxisomes are tiny, single membrane bound structures that contain enzymes which produce hydrogen peroxide as a by-product. These structures are involved in plant processes such as photorespiration.

- Plasmodesmata: These pores or channels are found between plant cell walls and allow molecules and communication signals to pass between individual plant cells.

- Ribosomes: Consisting of RNA and proteins, ribosomes are responsible for protein assembly. They can be found either attached to the rough ER or free in the cytoplasm.

- Vacuole: This plant cell organelle provides support for and participates in a variety of cellular functions including storage, detoxification, protection, and growth. When a plant cell matures, it typically contains one large liquid-filled vacuole.

Plant Cell Types

This is a typical dicotyledon stem (Buttercup). At center is an oval vascular
bundle embedded in parenchyma cells (yellow) of the cortex of the stem.
Some parenchyma cells contain chloroplasts (green).

As a plant matures, its cells become specialized in order to perform certain functions necessary for survival. Some plant cells synthesize and store organic products, while others help to transport nutrients throughout the plant. Some examples of specialized plant cell types and tissues include: parenchyma cells, collenchyma cells, sclerenchyma cells, xylem, and phloem.

Parenchyma Cells

Parenchyma cells are usually depicted as the typical plant cell because they are not as specialized as other cells. Parenchyma cells have thin walls and are found in dermal, ground, and vascular tissue systems. These cells help to synthesize and store organic products in the plant. The middle tissue layer of leaves (mesophyll) is composed of parenchyma cells, and it is this layer that contains plant chloroplasts.

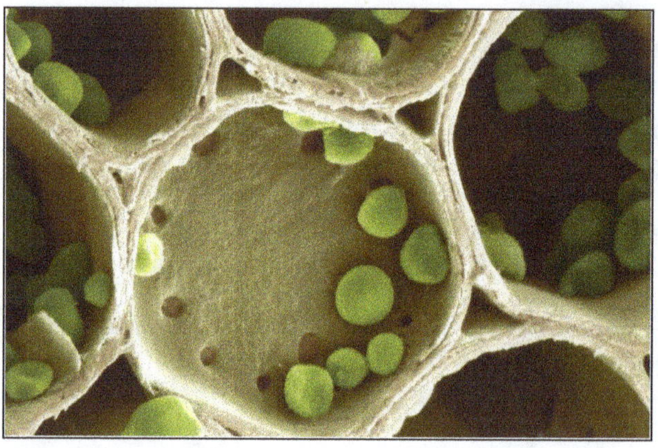

This image shows starch grains (green) in the parenchyma of a Clematis sp. plant. Starch is synthesized from the carbohydrate sucrose, a sugar produced by the plant during photosynthesis, and used as a source of energy. It is stored as grains in structures called amyloplasts (yellow).

Chloroplasts are plant organelles that are responsible for photosynthesis and most of the plant's metabolism takes place in parenchyma cells. Excess nutrients, often in the form of starch grains, are also stored in these cells. Parenchyma cells are not only found in plant leaves, but in the outer and inner layers of stems and roots as well. They are located between xylem and phloem and assist in the exchange of water, minerals, and nutrients. Parenchyma cells are the main components of plant ground tissue and the soft tissue of fruits.

Collenchyma Cells

Collenchyma cells have a support function in plants, particularly in young plants. These cells help to support plants, while not restraining growth. Collenchyma cells are elongated in shape and have thick primary cell walls composed of the carbohydrate polymers cellulose and pectin.

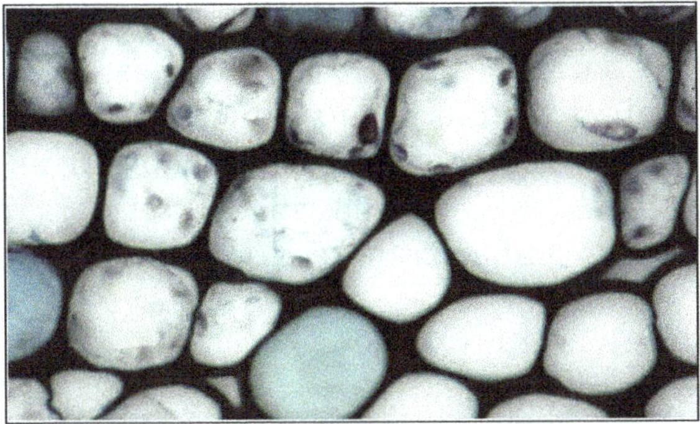

These plant collenchyma cells form supporting tissue.

Due to their lack of secondary cell walls and the absence of a hardening agent in their primary cell walls, collenchyma cells can provide structural support for tissues while maintaining flexibility. They are able to stretch along with a plant as it grows. Collenchyma cells are found in the cortex (layer between the epidermis and vascular tissue) of stems and along leaf veins.

Sclerenchyma Cells

Sclerenchyma cells also have a support function in plants, but unlike collenchyma cells, they have a hardening agent in their cell walls and are much more rigid. These cells have thick secondary cell walls and are non-living once matured. There are two types of sclerenchyma cells: sclereids and fibers.

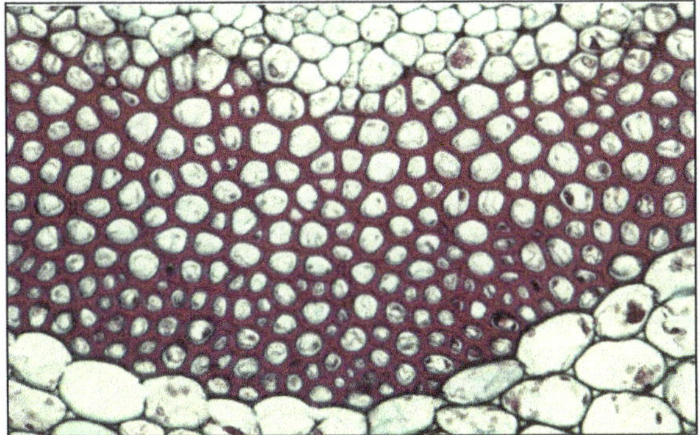

This images shows sclerenchyma at vascular bundles of a sunflower stem.

Sclerids have varied sizes and shapes, and most of the volume of these cells is taken up by the cell wall. Sclerids are very hard and form the hard outer shell of nuts and seeds. Fibers are elongated, slender cells that are strand-like in appearance. Fibers are strong and flexible and are found in stems, roots, fruit walls, and leaf vascular bundles.

Conducting Cells - Xylem and Phloem

Water conducting cells of xylem have a support function in plants. Xylem has a hardening agent in the tissue that makes it rigid and capable of functioning in structural support and transportation. The main function of xylem is to transport water throughout the plant. Two types of narrow, elongated cells compose xylem: tracheids and vessel elements. Tracheids have hardened secondary cell walls and function in water conduction. Vessel elements resemble open-ended tubes that are arranged end to end allowing water to flow within the tubes. Gymnosperms and seedless vascular plants contain tracheids, while angiosperms contain both tracheids and vessel members.

The center of this stem is filled with large xylem vessels for transporting water and mineral nutrients from the roots to the main body of the plant. Five bundles of phloem tissue (pale green) serve to distribute carbohydrate and plant hormones around the plant.

Vascular plants also have another type of conducting tissue called phloem. Sieve tube elements are the conducting cells of phloem. They transport organic nutrients, such as glucose, throughout the plant. The cells of sieve tube elements have few organelles allowing for easier passage of nutrients. Since sieve tube elements lack organelles, such as ribosomes and vacuoles, specialized parenchyma cells, called companion cells, must carry out metabolic functions for sieve tube elements. Phloem also contains sclerenchyma cells that provide structural support by increasing rigidity and flexibility.

Nuclear Envelope

The nuclear envelope is the outer covering of the nucleus in plant and other eukaryotic cells that acts as a barrier separating the nuclear contents from the surrounding cytoplasm.

The nuclear envelope is a double membrane system, consisting of two concentric membranes. The membranes are separated by a fluid-filled space called the perinuclear cisterna that measures about 20 to 40 nanometers. Like other plant cell membranes, the nuclear envelope consists of two bilayers, both made of phospholipids, in which numerous proteins are embedded.

Attachment sites for protein filaments are stitched on the innermost surface of the nuclear envelope. These protein filaments anchor the molecules of deoxyribonucleic acid (DNA) to the envelope and help to keep them organized. The network of filaments that enmesh the nuclear envelope provides stability.

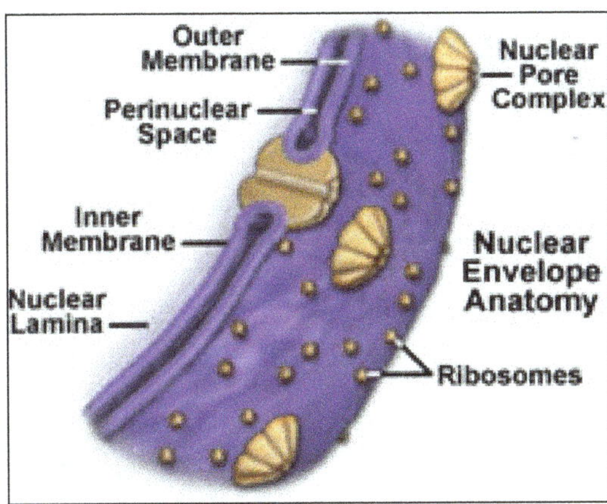

Large numbers of ribosomes are located on the outer surface of the envelope. The outer most membrane is continuous with the organelle called rough endoplasmic reticulum (RER) in the cytoplasm of the plant cell, which also has ribosomes attached to it.

The space between the outer and innermembranes is also continuous with the rough endoplasmic reticulum space and can fill with newly synthesized proteins, just as the RER does.When the nuclear envelope breaks down during cell division, its fragments are similar to portions of the endoplasmic reticulum.

This can be commonly observed at the root and shoot a pices of a plant, where active mitosis (regular cell division for growth) occurs. Both bilayers of the envelope are fused at intervals to form many nuclear pores, which consist of protein complexes in clusters. The two membranes enclose a flattened sac and are connected at the nuclear pore sites.

Functions

The nuclear envelope surrounds the fluid portion of the nucleus, called the nucleoplasm, in all plant cells. The ribosomes on the outer surface of the envelope serve as sites for protein synthesis in addition to ribosomes located in the cytoplasm.

The envelope is selectively permeable and therefore regulates the passage of materials and energy between the nucleoplasmand the cytoplasm. The envelope allows certain cell activities to be localized within the nucleus or outside in the cytoplasm. It also permits many different activities to go on simultaneously within and outside of the nucleus.

Like other membranes of the plant cell, the nuclear envelope membranes serve as important work surfaces for many chemical reactions in plants that are carried out by enzymes bound to the membranes. These functions are essential in the plant for the transport and use of minerals and water from the roots to cells of the stem and leaves.

They are also important for the movement and use of carbohydrates, proteins, lipids, nucleic acids, and other chemical compounds in plant cells after photosynthesis, protein synthesis, and other biochemical activities have occurred.

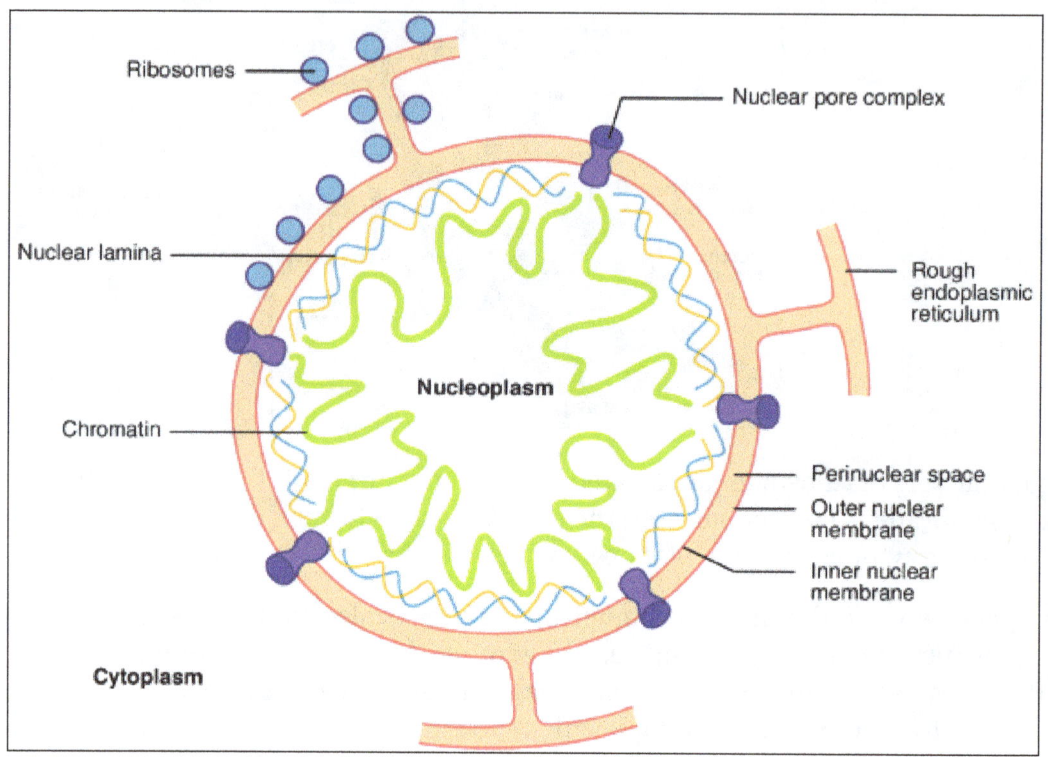

Nuclear Pores

Nuclear pores, of about 100 nanometers in diameter, perforate the nuclear envelope. These pores look like wheels with eight spokes when observed from the top. Each contains eight subunits over the region where the inner and outer membranes join. They form a ring of subunits that are 15 to 20 nanometers in diameter. At the lip of each pore, the inner and outer membranes of the nuclear envelope are fused.

Each nuclear pore serves as a water-filled channel, and the arrangement allows transport in and out of the nucleus to occur in several ways. The nuclear pores allow the passage of materials to the cytoplasm from the interior of the nucleus, and vice versa, but the process is highly selective, permitting only specific molecules to pass through these openings.

An intricate protein structure called the pore complex lines each pore and regulates the entry and exit of certain large macromolecules and particles. Large molecules, including the subunits of ribosomes, cross the bilayers at the pores in highly controlled ways. Ions and small, water-soluble molecules cross the nuclear envelope at the pores.

Studies show that the pore can actually dilate more when it gets the appropriate signal. Studies have also shown that the signal is in the peptide sequences of themolecules.

These signals are recognition sequences rich in the amino acids lysine, arginine, and proline.

Nuclear pores in plants therefore exert control over the movement of materials. This is, for example, demonstrated in the fact that if a nucleus is extracted from a cell and placed into water, it swells; this can happen only if the pores prevent material from oozing out as the nucleus absorbs water.

Nuclear Lamina

The nuclear lamina is a layer of specific proteins, called lamins, attached to inside membrane of the nuclear envelope. The layer consists of thin filaments (intermediate filaments) that are 30 to 40 nanometers thick.

Each filament is a polymer of lamin. There are two types of lamin: A-type lamins are inside, next to the nucleoplasm, and B-type lamins are near the inner part of the nuclear membrane.

The nuclear lamina surrounds the nucleus, except at the nuclear pores. The lamina serves as a skeletal framework for the nucleus. It may be involved in the functional organization of the nucleus and may also play an important role in the breakdown and reassembly of the nuclear envelope during mitosis.

Cell Wall

Cell wall is a specialized form of extracellular matrix that surrounds every cell of a plant. The cell wall is responsible for many of the characteristics that distinguish plant cells from animal cells. Although often perceived as an inactive product serving mainly mechanical and structural purposes, the cell wall actually has a multitude of functions upon which plant life depends. Such functions include: (1) providing the living cell with mechanical protection and a chemically buffered environment, (2) providing a porous medium for the circulation and distribution of water, minerals, and other small nutrient molecules, (3) providing rigid building blocks from which stable structures of higher order, such as leaves and stems, can be produced, and (4) providing a storage site of regulatory molecules that sense the presence of pathogenic microbes and control the development of tissues.

Certain prokaryotes, algae, slime molds, water molds, and fungi also have cell walls. Bacterial cell walls are characterized by the presence of peptidoglycan, whereas those of Archaea characteristically lack this chemical. Algal cell walls are similar to those of plants, and many contain specific polysaccharides that are useful for taxonomy. Unlike those of plants and algae, fungal cell walls lack cellulose entirely and contain chitin.

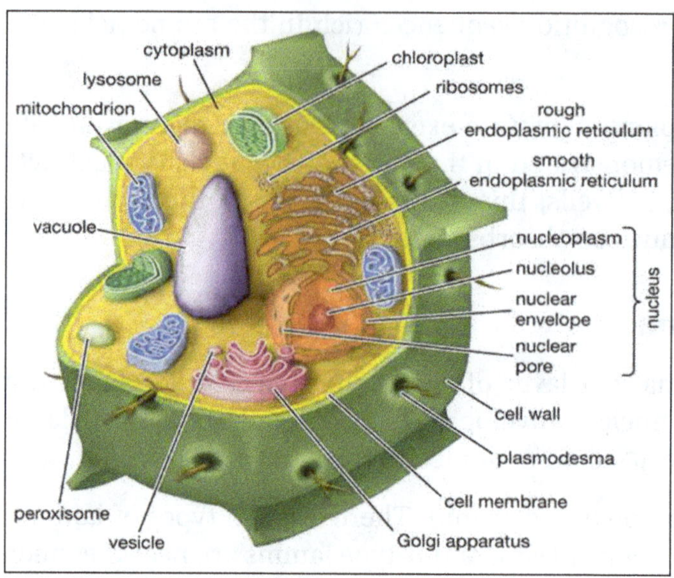

Cutaway drawing of a plant cell, showing the cell wall and internal organelles.

Mechanical Properties

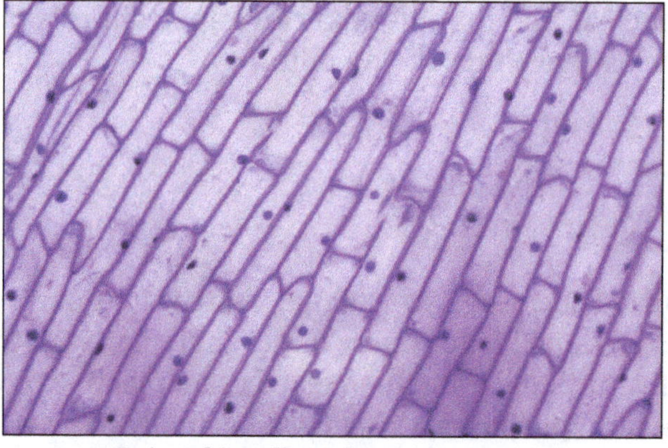

Onion skin cells under a microscope.

All cell walls contain two layers, the middle lamella and the primary cell wall, and many cells produce an additional layer, called the secondary wall. The middle lamella serves as a cementing layer between the primary walls of adjacent cells. The primary wall is the cellulose-containing layer laid down by cells that are dividing and growing. To allow for cell wall expansion during growth, primary walls are thinner and less rigid than those of cells that have stopped growing. A fully grown plant cell may retain its primary cell wall (sometimes thickening it), or it may deposit an additional, rigidifying layer of different composition, which is the secondary cell wall. Secondary cell walls are responsible for most of the plant's mechanical support as well as the mechanical properties prized in wood. In contrast to the permanent stiffness and load-bearing capacity of thick secondary walls, the thin primary walls are capable of serving a structural, supportive role only when the vacuoles within the cell are filled with water to the point that they exert a turgor pressure against the cell wall. Turgor-induced stiffening of primary walls is analogous to the stiffening of the sides of a pneumatic tire by air pressure. The wilting of flowers and leaves is caused by a loss of turgor pressure, which results in turn from the loss of water from the plant cells.

Components

Although primary and secondary wall layers differ in detailed chemical composition and structural organization, their basic architecture is the same, consisting of cellulose fibres of great tensile strength embedded in a water-saturated matrix of polysaccharides and structural glycoproteins.

Cellulose

Cellulose consists of several thousand glucose molecules linked end to end. The chemical links between the individual glucose subunits give each cellulose molecule a flat ribbonlike structure that allows adjacent molecules to band laterally together into microfibrils with lengths ranging from two to seven micrometres. Cellulose fibrils are synthesized by enzymes floating in the cell membrane and are arranged in a rosette configuration. Each rosette appears capable of "spinning" a microfibril into the cell wall. During this process, as new glucose subunits are added to the growing end of the fibril, the rosette is pushed around the cell on the surface of the cell membrane, and its cellulose fibril becomes wrapped around the protoplast. Thus, each plant cell can be viewed as making its own cellulose fibril cocoon.

Cellulose consists of glucose molecules linked end to end.

Matrix Polysaccharides

The two major classes of cell wall matrix polysaccharides are the hemicelluloses and the pectic polysaccharides, or pectins. Both are synthesized in the Golgi apparatus, brought to the cell surface in small vesicles, and secreted into the cell wall.

Hemicelluloses consist of glucose molecules arranged end to end as in cellulose, with short side chains of xylose and other uncharged sugars attached to one side of the ribbon. The other side of the ribbon binds tightly to the surface of cellulose fibrils, thereby coating the microfibrils with hemicellulose and preventing them from adhering together in an uncontrolled manner. Hemicellulose molecules have been shown to regulate the rate at which primary cell walls expand during growth.

The heterogeneous, branched, and highly hydrated pectic polysaccharides differ from hemicelluloses in important respects. Most notably, they are negatively charged because of galacturonic acid residues, which, together with rhamnose sugar molecules, form the linear backbone of all pectic polysaccharides. The backbone contains stretches of pure galacturonic acid residues interrupted by segments in which galacturonic acid and rhamnose residues alternate; attached to these latter segments are complex, branched sugar side chains. Because of their negative charge, pectic polysaccharides bind tightly to positively charged ions, or cations. In cell walls, calcium ions cross-link the stretches of pure galacturonic acid residues tightly, while leaving the rhamnose-containing segments in a more open, porous configuration. This cross-linking creates the semirigid gel properties characteristic of the cell wall matrix—a process exploited in the preparation of jellied preserves.

Proteins

Although plant cell walls contain only small amounts of protein, they serve a number of important functions. The most prominent group are the hydroxyproline-rich glycoproteins, shaped like rods with connector sites, of which extensin is a prominent example. Extensin contains 45 percent hydroxyproline and 14 percent serine residues distributed along its length. Every hydroxyproline residue carries a short side chain of arabinose sugars, and most serine residues carry a galactose sugar. This gives rise to long molecules, resembling bottle brushes, that are secreted into the cell wall toward the end of primary wall formation and become covalently cross-linked into a mesh at the time that cell growth stops. Plant cells may control their ultimate size by regulating the time at which this cross-linking of extensin molecules occurs.

In addition to the structural proteins, cell walls contain a variety of enzymes. Most notable are those that cross-link extensin, lignin, cutin, and suberin molecules into networks. Other enzymes help protect plants against fungal pathogens by breaking fragments off of the cell walls of the fungi. The fragments in turn induce defense responses in underlying cells. The softening of ripe fruit and dropping of leaves in the autumn are brought about by cell wall-degrading enzymes.

Plastics

Cell wall plastics such as lignin, cutin, and suberin all contain a variety of organic compounds cross-linked into tight three-dimensional networks that strengthen cell walls and make them more resistant to fungal and bacterial attack. Lignin is the general name for a diverse group of polymers of aromatic alcohols. Deposited mostly in secondary cell walls and providing the rigidity of terrestrial vascular plants, it accounts for up to 30 percent of a plant's dry weight. The diversity of cross-links between the polymers—and the resulting tightness—makes lignin a formidable barrier to the penetration of most microbes.

Cutin and suberin are complex biopolyesters composed of fatty acids and aromatic compounds. Cutin is the major component of the cuticle, the waxy, water-repelling surface layer of cell walls exposed to the environment aboveground. By reducing the wettability of leaves and stems—and thereby affecting the ability of fungal spores to germinate—it plays an important part in the defense strategy of plants. Suberin serves with waxes as a surface barrier of underground parts. Its synthesis is also stimulated in cells close to wounds, thereby sealing off the wound surfaces and protecting underlying cells from dehydration.

Intercellular Communication

Plasmodesmata

Similar to the gap junction of animal cells is the plasmodesma, a channel passing through the cell wall and allowing direct molecular communication between adjacent plant cells. Plasmodesmata are lined with cell membrane, in effect uniting all connected cells with one continuous cell membrane. Running down the middle of each channel is a thin membranous tube that connects the endoplasmic reticula (ER) of the two cells. This structure is a remnant of the ER of the original parent cell, which, as the parent cell divided, was caught in the developing cell plate.

Although the precise mechanisms are not fully understood, the plasmodesma is thought to regulate the passage of small molecules such as salts, sugars, and amino acids by constricting or dilating the openings at each end of the channel.

Oligosaccharides with Regulatory Functions

The discovery of cell wall fragments with regulatory functions opened a new era in plant research. For years scientists had been puzzled by the chemical complexity of cell wall polysaccharides, which far exceeds the structural requirements of plant cell walls. The answer came when it was found that specific fragments of cell wall polysaccharides, called oligosaccharins, are able to induce specific responses in plant cells and tissues. One such fragment, released by enzymes used by fungi to break down plant cell walls, consists of a linear polymer of 10 to 12 galacturonic acid residues. Exposure of plant cells to such fragments induces them to produce antibiotics known as phytoalexins. In other experiments it has been shown that exposing strips of tobacco stem cells to a different type of cell wall fragment leads to the growth of roots; other fragments lead to the formation of stems and yet others to the production of flowers. In all instances the concentration of oligosaccharins required to bring about the observed responses is equal to that of hormones in animal cells; indeed, oligosaccharins may be viewed as the oligosaccharide hormones of plants.

Central Vacuole

The central vacuole is a large vacuole found inside of plant cells. A vacuole is a sphere filled with fluid and molecules inside a cell. The central vacuole stores water and maintains turgor pressure in a plant cell. It also pushes the contents of the cell toward the cell membrane, which allows the plant cells to take in more light energy for making food through photosynthesis. Vacuoles are also found in animal, protist, fungal, and bacterial cells, but large central vacuoles are only found in plant cells.

Function of the Central Vacuole

The main function of the central vacuole is to maintain turgor pressure in the cell. Turgor pressure is the pressure of the cell's contents pushing against the cell wall; it is only found in cells that have cell walls, such as those of plants, fungi, and bacteria. Turgor pressure changes in a cell due to osmosis, which is the diffusion of water into or out of the cell. When a plant cell is in a hypotonic solution, there is a higher concentration of water molecules outside the cell than inside, and water will flow into the cell. In plants, this causes the vacuole to be filled with water, and the cell has high turgidity. This is the optimal condition for plant cells. Isotonic solutions have roughly the same concentration of water molecules within and outside of the cell membrane, so the amount of water leaving and entering is the same. Plant cells become flaccid in isotonic solutions, and the plant may start to droop. In hypertonic solutions, where there is more water inside the cell than outside, water will flow out of the cell, and the plant will wilt and possibly die. The central vacuole is able to store a lot of water and swell up so that plant cells can maintain the high turgidity needed for the plant to function optimally. The central vacuole can also store waste products and nutrients temporarily, and the concentration of these also affects turgor pressure; having molecules other than water in the central vacuole decreases turgidity, so the cell must always have a much higher concentration of water in its central vacuole than any other molecule.

Plant cells thrive in hypotonic solutions because their cell walls keep them from bursting due to excessive water intake. By contrast, animal cells, which do not have cell walls, fare best in isotonic solutions. If animal cells are in hypotonic solutions, too much water will enter the cell and the cell can burst.

The central vacuole can take up anywhere from 30-90 percent of a plant cell's space inside the cell membrane, and one of the central vacuole's other functions is to push the other contents of the cell closer to the cell membrane. This allows organelles inside the plant cell called chloroplasts to receive more light, which is very important because photosynthesis occurs in chloroplasts. Photosynthesis is the production of nutrients from light energy, carbon dioxide, and water; it is how a plant makes its own food. By pushing chloroplasts closer to the surface of the cell, the central vacuole makes it possible for chloroplasts to take in more energy from sunlight.

Structure of the Central Vacuole

The central vacuole consists of two parts, the cell sap and the tonoplast. The cell sap refers to the fluid within the vacuole. It is mostly water, but also consists of ions, salts, waste products, nutrients, and sometimes pigment molecules. The tonoplast is the central vacuole's membrane; it is also known as the vacuolar membrane. It separates the contents of the central vacuole from the rest of the cell. It is made up of phospholipids and proteins, just like the cell membrane that covers the plant cell. The proteins in the membrane of the tonoplast can control water entry and exit in the central vacuole along with regulating the movements of ions such as potassium.

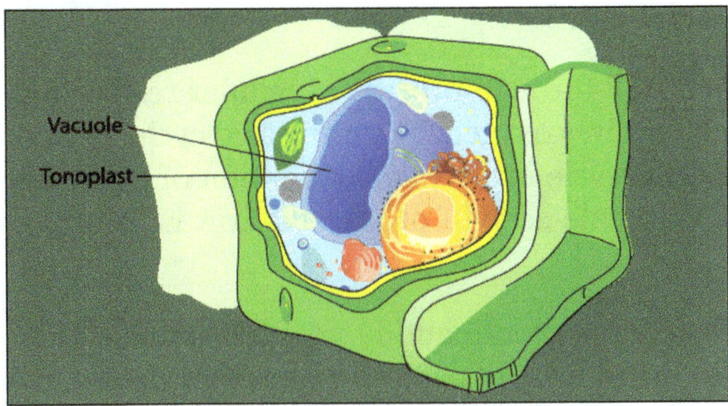

Cell Nucleus

The nucleus is a highly specialized organelle that serves as the information and administrative center of the cell. This organelle has two major functions. It stores the cell's hereditary material, or DNA, and it coordinates the cell's activities, which include intermediary metabolism, growth, protein synthesis, and reproduction (cell division).

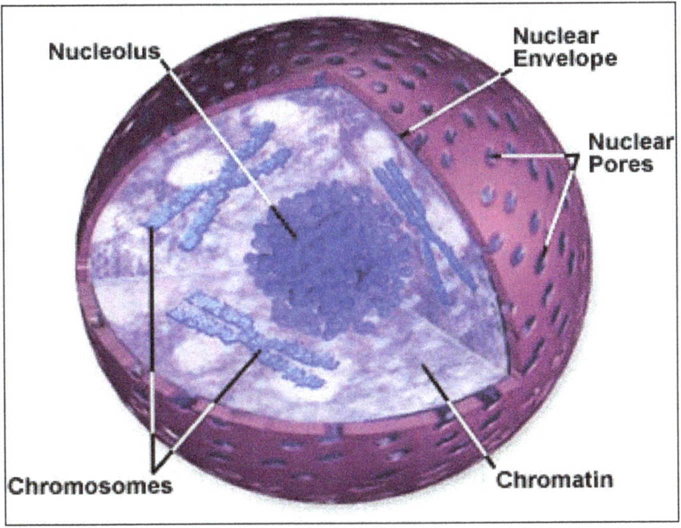

Mitochondria

Mitochondria (singular: mitochondrion) are organelles within eukaryotic cells that produce adenosine triphosphate (ATP), the main energy molecule used by the cell. For this reason, the mitochondrion is sometimes referred to as "the powerhouse of the cell". Mitochondria are found in all eukaryotes, which are all living things that are not bacteria or archaea. It is thought that mitochondria arose from once free-living bacteria that were incorporated into cells.

Function of Mitochondria

Mitochondria produce ATP through process of cellular respiration—specifically, aerobic respiration, which requires oxygen. The citric acid cycle, or Krebs cycle, takes place in the mitochondria. This cycle involves the oxidation of pyruvate, which comes from glucose, to form the molecule acetyl-CoA. Acetyl-CoA is in turn oxidized and ATP is produced.

The citric acid cycle reduces nicotinamide adenine dinucleotide (NAD^+) to NADH. NADH is then used in the process of oxidative phosphorylation, which also takes place in the mitochondria. Electrons from NADH travel through protein complexes that are embedded in the inner membrane of the mitochondria. This set of proteins is called an electron transport chain. Energy from the electron transport chain is then used to transport proteins back across the membrane, which power ATP synthase to form ATP.

The amount of mitochondria in a cell depends on how much energy that cell needs to produce. Muscle cells, for example, have many mitochondria because they need to produce energy to move the body. Red blood cells, which carry oxygen to other cells, have none; they do not need to produce energy. Mitochondria are analogous to a furnace or a powerhouse in the cell because, like furnaces and powerhouses, mitochondria produce energy from basic components (in this case, molecules that have been broken down so that they can be used).

Mitochondria have many other functions as well. They can store calcium, which maintains homeostasis of calcium levels in the cell. They also regulate the cell's metabolism and have roles in apoptosis (controlled cell death), cell signaling, and thermogenesis (heat production).

Structure of Mitochondria

Mitochondria have two membranes, an outer membrane and an inner membrane. These membranes are made of phospholipid layers, just like the cell's outer membrane. The outer membrane covers the surface of the mitochondrion, while the inner membrane is located within and has many folds called cristae. The folds increase surface area of the membrane, which is important because the inner membrane holds the proteins involved in the electron transport chain. It is also where many other chemical reactions take place to carry out the mitochondria's many functions. An increased surface area creates more space for more reactions to occur, and increases the mitochondria's output. The space between the outer and inner membranes is called the intermembrane space, and the space inside the inner membrane is called the matrix.

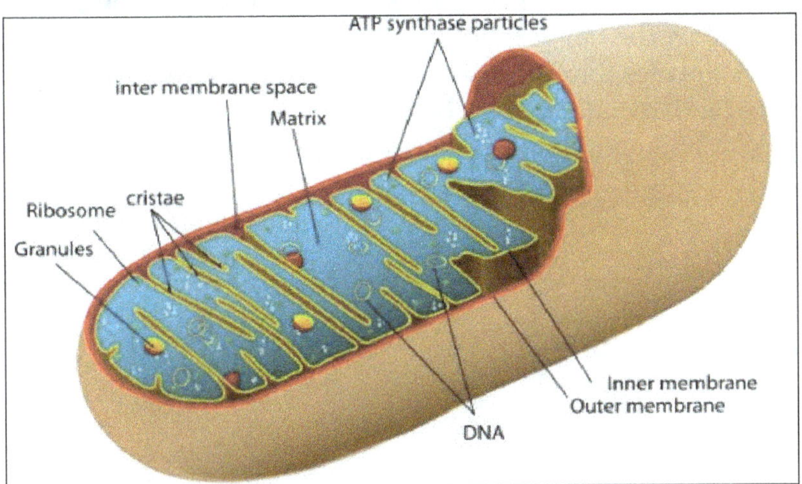

This diagram shows the structure of a mitochondrion.

Plastid

The plastid is a membrane-bound organelle found in the cells of plants, algae, and some other eukaryotic organisms. They are considered endosymbiotic Cyanobacteria, related to the Gloeomargarita. Plastids were discovered and named by Ernst Haeckel, but A. F. W. Schimper was the first to provide a clear definition. Plastids are the site of manufacture and storage of important chemical compounds used by the cells of autotrophic eukaryotes. They often contain pigments used in photosynthesis, and the types of pigments in a plastid determine the cell's color. They have a common evolutionary origin and possess a double-stranded DNA molecule that is circular, like that of prokaryotic cells.

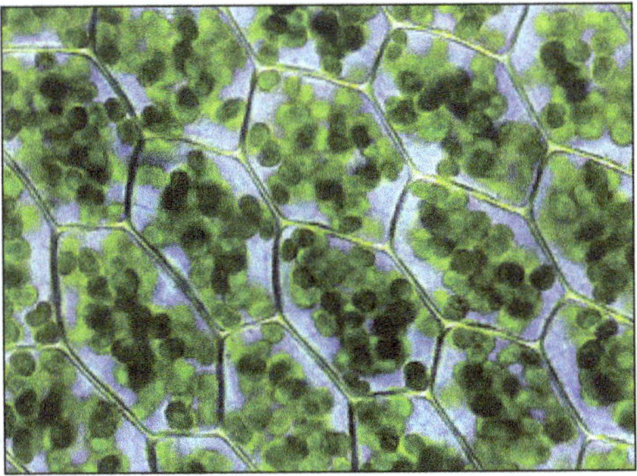

Plant cells with visible chloroplasts.

In Plants

Plastids that contain chlorophyll can carry out photosynthesis and are called chloroplasts. Plastids can also store products like starch and can synthesize fatty acids and terpenes, which can be used for producing energy and as raw material for the synthesis of other molecules. For example, the components of the plant cuticle and its epicuticular wax are synthesized by the epidermal cells from palmitic acid, which is synthesized in the chloroplasts of the mesophyll tissue. All plastids are derived from proplastids, which are present in the meristematic regions of the plant. Proplastids and young chloroplasts commonly divide by binary fission, but more mature chloroplasts also have this capacity.

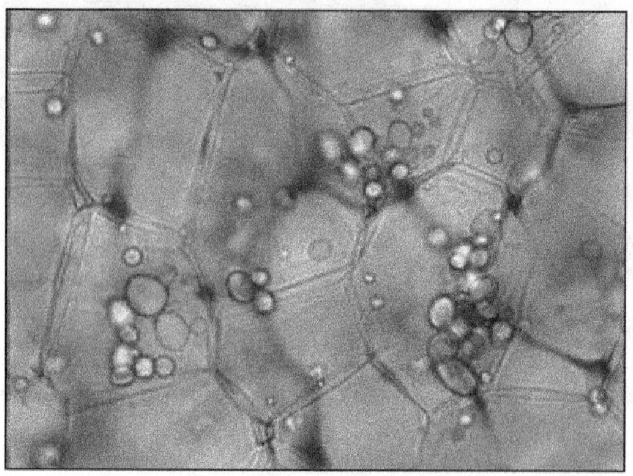

Leucoplasts in plant cells.

In plants, plastids may differentiate into several forms, depending upon which function they perform in the cell. Undifferentiated plastids (proplastids) may develop into any of the following variants:

- Chloroplasts: Green plastids for photosynthesis.

- Chromoplasts: Coloured plastids for pigment synthesis and storage.

- Gerontoplasts: Control the dismantling of the photosynthetic apparatus during plant senescence.

- Leucoplasts: Colourless plastids for monoterpene synthesis; leucoplasts sometimes differentiate into more specialized plastids:

  - Amyloplasts: For starch storage and detecting gravity (for geotropism).

  - Elaioplasts: For storing fat.

  - Proteinoplasts: For storing and modifying protein.

  - Tannosomes: For synthesizing and producing tannins and polyphenols.

Depending on their morphology and function, plastids have the ability to differentiate, or redifferentiate, between these and other forms.

Each plastid creates multiple copies of a circular 75–250 kilobase plastome. The number of genome copies per plastid is variable, ranging from more than 1000 in rapidly dividing cells, which, in general, contain few plastids, to 100 or fewer in mature cells, where plastid divisions have given rise to a large number of plastids. The plastome contains about 100 genes encoding ribosomal and transfer ribonucleic acids (rRNAs and tRNAs) as well as proteins involved in photosynthesis and plastid gene transcription and translation. However, these proteins only represent a small fraction of the total protein set-up necessary to build and maintain the structure and function of a particular type of plastid. Plant nuclear genes encode the vast majority of plastid proteins, and the expression of plastid genes and nuclear genes is tightly co-regulated to coordinate proper development of plastids in relation to cell differentiation.

Plastid DNA exists as large protein-DNA complexes associated with the inner envelope membrane and called 'plastid nucleoids'. Each nucleoid particle may contain more than 10 copies of the plastid DNA. The proplastid contains a single nucleoid located in the centre of the plastid. The developing plastid has many nucleoids, localized at the periphery of the plastid, bound to the inner envelope membrane. During the development of proplastids to chloroplasts, and when plastids convert from one type to another, nucleoids change in morphology, size and location within the organelle. The remodelling of nucleoids is believed to occur by modifications to the composition and abundance of nucleoid proteins.

Many plastids, particularly those responsible for photosynthesis, possess numerous internal membrane layers.

In plant cells, long thin protuberances called stromules sometimes form and extend from the main plastid body into the cytosol and interconnect several plastids. Proteins, and presumably smaller molecules, can move within stromules. Most cultured cells that are relatively large compared to other plant cells have very long and abundant stromules that extend to the cell periphery.

In 2014, evidence of possible plastid genome loss was found in Rafflesia lagascae, a non-photosynthetic parasitic flowering plant, and in Polytomella, a genus of non-photosynthetic green algae. Extensive searches for plastid genes in both Rafflesia and Polytomella yielded no results, however the conclusion that their plastomes are entirely missing is still controversial. Some scientists argue that plastid genome loss is unlikely since even non-photosynthetic plastids contain genes necessary to complete various biosynthetic pathways, such as heme biosynthesis.

In Algae

In algae, the term leucoplast is used for all unpigmented plastids. Their function differs from the leucoplasts of plants. Etioplasts, amyloplasts and chromoplasts are plant-specific and do not occur in algae. Plastids in algae and hornworts may also differ from plant plastids in that they contain pyrenoids.

Glaucophyte algae contain muroplasts, which are similar to chloroplasts except that they have a peptidoglycan cell wall that is similar to that of prokaryotes. Red algae contain rhodoplasts, which are red chloroplasts that allow them to photosynthesize to a depth of up to 268 m. The chloroplasts of plants differ from the rhodoplasts of red algae in their ability to synthesize starch, which is stored in the form of granules within the plastids. In red algae, floridean starch is synthesized and stored outside the plastids in the cytosol.

Inheritance

Most plants inherit the plastids from only one parent. In general, angiosperms inherit plastids from the female gamete, whereas many gymnosperms inherit plastids from the male pollen. Algae also inherit plastids from only one parent. The plastid DNA of the other parent is, thus, completely lost.

In normal intraspecific crossings (resulting in normal hybrids of one species), the inheritance of plastid DNA appears to be quite strictly 100% uniparental. In interspecific hybridisations, however, the inheritance of plastids appears to be more erratic. Although plastids inherit mainly maternally in interspecific hybridisations, there are many reports of hybrids of flowering plants that contain plastids of the father. Approximately 20% of angiosperms, including alfalfa (Medicago sativa), normally show biparental inheritance of plastids.

DNA Damage and Repair

Plastid DNA of maize seedlings is subject to increased damage as the seedlings develop. The DNA is damaged in oxidative environments created by photo-oxidative reactions and photosynthetic/respiratory electron transfer. Some DNA molecules are repaired while DNA with unrepaired damage appears to be degraded to non-functional fragments.

DNA repair proteins are encoded by the cell's nuclear genome but can be translocated to plastids where they maintain genome stability/integrity by repairing the plastid's DNA. As an example, in chloroplasts of the moss Physcomitrella patens, a protein employed in DNA mismatch repair (Msh1) interacts with proteins employed in recombinational repair (RecA and RecG) to maintain plastid genome stability.

Chloroplast

Chloroplasts are organelles that conduct photosynthesis, where the photosynthetic pigment chlorophyll captures the energy from sunlight, converts it, and stores it in the energy-storage molecules ATP and NADPH while freeing oxygen from water in plant and algal cells. They then use the ATP and NADPH to make organic molecules from carbon dioxide in a process known as the Calvin cycle. Chloroplasts carry out a number of other functions, including fatty acid synthesis, much amino acid synthesis, and the immune response in plants. The number of chloroplasts per cell varies from one, in unicellular algae, up to 100 in plants like Arabidopsis and wheat.

Chloroplasts visible in the cells of Bryum capillare, a type of moss.

A chloroplast is a type of organelle known as a plastid, characterized by its two membranes and a high concentration of chlorophyll. Other plastid types, such as the leucoplast and the chromoplast, contain little chlorophyll and do not carry out photosynthesis.

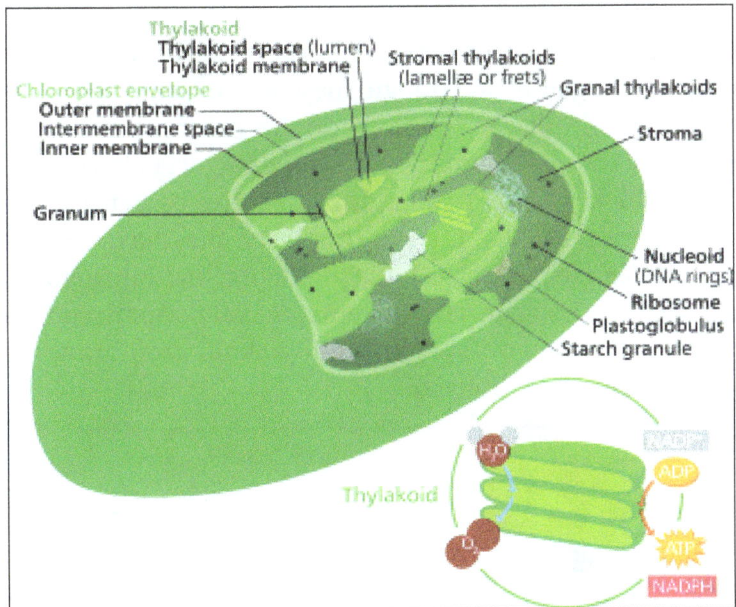

Structure of a typical higher-plant chloroplast.

Chloroplasts are highly dynamic—they circulate and are moved around within plant cells, and occasionally pinch in two to reproduce. Their behavior is strongly influenced by environmental factors like light color and intensity. Chloroplasts, like mitochondria, contain their own DNA, which is thought to be inherited from their ancestor—a photosynthetic cyanobacterium that was engulfed by an early eukaryotic cell. Chloroplasts cannot be made by the plant cell and must be inherited by each daughter cell during cell division.

With one exception (the amoeboid Paulinella chromatophora), all chloroplasts can probably be traced back to a single endosymbiotic event, when a cyanobacterium was engulfed by the eukaryote. Despite this, chloroplasts can be found in an extremely wide set of organisms, some not even directly related to each other—a consequence of many secondary and even tertiary endosymbiotic events.

Lineages and Evolution

Chloroplasts are one of many types of organelles in the plant cell. They are considered to have evolved from endosymbiotic cyanobacteria. Mitochondria are thought to have come from a similar endosymbiosis event, where an aerobic prokaryote was engulfed. This origin of chloroplasts was first suggested by the Russian biologist Konstantin Mereschkowski in 1905 after Andreas Schimper observed in 1883 that chloroplasts closely resemble cyanobacteria. Chloroplasts are only found in plants, algae,and the amoeboid Paulinella chromatophora.

Parent Group: Cyanobacteria

Chloroplasts are considered endosymbiotic Cyanobacteria. Cyanobacteria are sometimes called blue-green algae even though they are prokaryotes. They are a diverse phylum of bacteria capable of carrying out photosynthesis, and are gram-negative, meaning that they have two cell membranes. Cyanobacteria also contain a peptidoglycan cell wall, which is thicker than in other gram-negative bacteria, and which is located between their two cell membranes. Like chloroplasts, they have thylakoids within. On the thylakoid membranes are photosynthetic pigments, including chlorophyll a. Phycobilins are also common cyanobacterial pigments, usually organized into hemispherical phycobilisomes attached to the outside of the thylakoid membranes (phycobilins are not shared with all chloroplasts though).

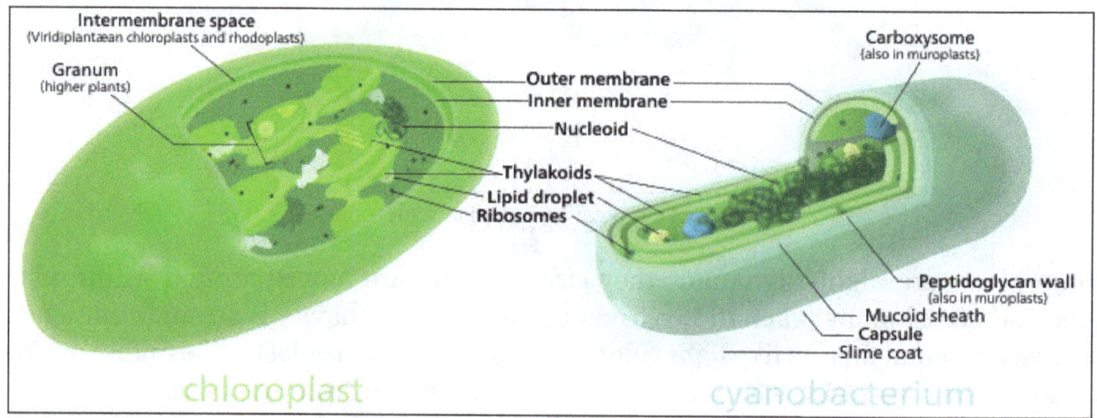

Both chloroplasts and cyanobacteria have a double membrane, DNA, ribosomes, and thylakoids. Both the chloroplast and cyanobacterium depicted are idealized versions (the chloroplast is that of a higher plant)—a lot of diversity exists among chloroplasts and cyanobacteria.

Primary Endosymbiosis

A eukaryote with mitochondria engulfed a cyanobacterium in an event of serial primary endosymbiosis, creating a lineage of cells with both organelles. It is important to note that the cyanobacterial endosymbiont already had a double membrane—the phagosomal vacuole-derived membrane was lost.

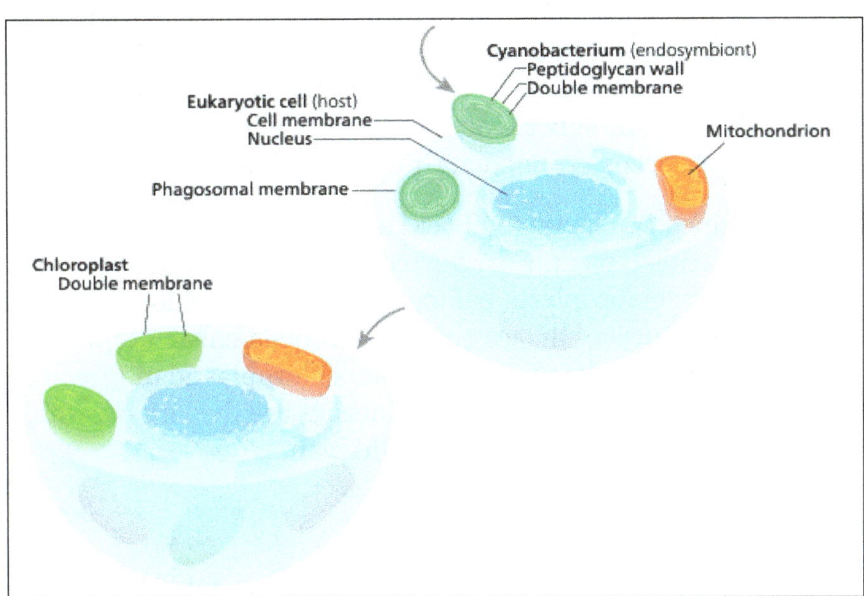

Somewhere around 1 to 2 billion years ago, a free-living cyanobacterium entered an early eukaryotic cell, either as food or as an internal parasite, but managed to escape the phagocytic vacuole it was contained in. The two innermost lipid-bilayer membranes that surround all chloroplasts correspond to the outer and inner membranes of the ancestral cyanobacterium's gram negative cell wall, and not the phagosomal membrane from the host, which was probably lost. The new cellular resident quickly became an advantage, providing food for the eukaryotic host, which allowed it to live within it. Over time, the cyanobacterium was assimilated, and many of its genes were lost or transferred to the nucleus of the host. From genomes that probably originally contained over 3000 genes only about 130 genes remain in the chloroplasts of contemporary plants. Some of its proteins were then synthesized in the cytoplasm of the host cell, and imported back into the chloroplast (formerly the cyanobacterium). Separately, somewhere around 500 million years ago, it happened again and led to the amoeboid Paulinella chromatophora.

This event is called endosymbiosis, or "cell living inside another cell with a mutual benefit for both". The external cell is commonly referred to as the host while the internal cell is called the endosymbiont.

Chloroplasts are believed to have arisen after mitochondria, since all eukaryotes contain mitochondria, but not all have chloroplasts. This is called serial endosymbiosis—an early eukaryote engulfing the mitochondrion ancestor, and some descendants of it then engulfing the chloroplast ancestor, creating a cell with both chloroplasts and mitochondria.

Whether or not primary chloroplasts came from a single endosymbiotic event, or many independent engulfments across various eukaryotic lineages, has long been debated. It is now generally held that organisms with primary chloroplasts share a single ancestor that took in a cyanobacterium 600–2000 million years ago. It has been proposed this the closest living relative of this bacterium is Gloeomargarita lithophora. The exception is the amoeboid Paulinella chromatophora, which descends from an ancestor that took in a Prochlorococcus cyanobacterium 90–500 million years ago.

These chloroplasts, which can be traced back directly to a cyanobacterial ancestor, are known as primary plastids ("plastid" in this context means almost the same thing as chloroplast). All primary chloroplasts belong to one of four chloroplast lineages—the glaucophyte chloroplast lineage, the amoeboid Paulinella chromatophora lineage, the rhodophyte (red algal) chloroplast lineage, or the chloroplastidan (green) chloroplast lineage. The rhodophyte and chloroplastidan lineages are the largest, with chloroplastidan (green) being the one that contains the land plants.

Glaucophyta

Usually the endosymbiosis event is considered to have occurred in the Archaeplastida, within which the glaucophyta being the possible earliest diverging lineage. The glaucophyte chloroplast group is the smallest of the three primary chloroplast lineages, being found in only 13 species, and is thought to be the one that branched off the earliest. Glaucophytes have chloroplasts that retain a peptidoglycan wall between their double membranes, like their cyanobacterial parent. For this reason, glaucophyte chloroplasts are also known as 'muroplasts' (besides 'cyanoplasts' or 'cyanelles'). Glaucophyte chloroplasts also contain concentric unstacked thylakoids, which surround a carboxysome – an icosahedral structure that glaucophyte chloroplasts and cyanobacteria keep their carbon fixation enzyme RuBisCO in. The starch that they synthesize collects outside the chloroplast. Like cyanobacteria, glaucophyte and rhodophyte chloroplast thylakoids are studded with light collecting structures called phycobilisomes. For these reasons, glaucophyte chloroplasts are considered a primitive intermediate between cyanobacteria and the more evolved chloroplasts in red algae and plants.

The rhodophyte, or red algal chloroplast group is another large and diverse chloroplast lineage. Rhodophyte chloroplasts are also called rhodoplasts, literally "red chloroplasts".

Rhodoplasts have a double membrane with an intermembrane space and phycobilin pigments organized into phycobilisomes on the thylakoid membranes, preventing their thylakoids from stacking. Some contain pyrenoids. Rhodoplasts have chlorophyll a and phycobilins for photosynthetic pigments; the phycobilin phycoerytherin is responsible for giving many red algae their distinctive red color. However, since they also contain the blue-green chlorophyll a and other pigments, many are reddish to purple from the combination. The red phycoerytherin pigment is an adaptation to help red algae catch more sunlight in deep water—as such, some red algae that live in shallow water have less phycoerytherin in their rhodoplasts, and can appear more greenish. Rhodoplasts synthesize a form of starch called floridean starch, which collects into granules outside the rhodoplast, in the cytoplasm of the red alga.

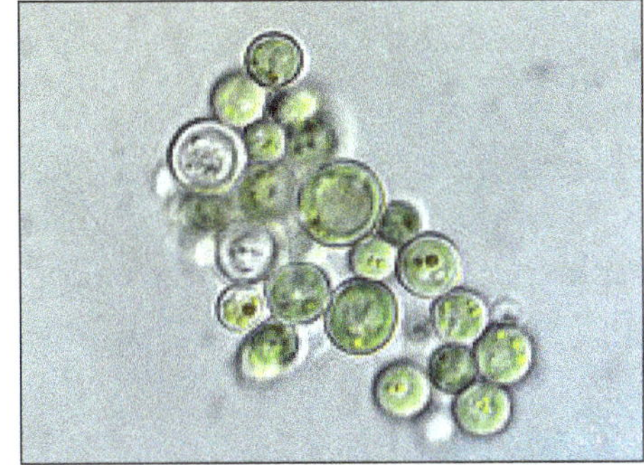

Diversity of red algae. Red algal chloroplasts are characterized by phycobilin pigments which often give them their reddish color.

Chloroplastida (Green Algae and Plants)

The chloroplastidan chloroplasts, or green chloroplasts, are another large, highly diverse primary chloroplast lineage. Their host organisms are commonly known as the green algae and land plants. They differ from glaucophyte and red algal chloroplasts in that they have lost their phycobilisomes, and contain chlorophyll b instead. Most green chloroplasts are (obviously) green, though some aren't, like some forms of Hæmatococcus pluvialis, due to accessory pigments that override the chlorophylls' green colors. Chloroplastidan chloroplasts have lost the peptidoglycan wall between their double membrane, leaving an intermembrane space. Some plants seem to have kept the genes for the synthesis of the peptidoglycan layer, though they've been repurposed for use in chloroplast division instead.

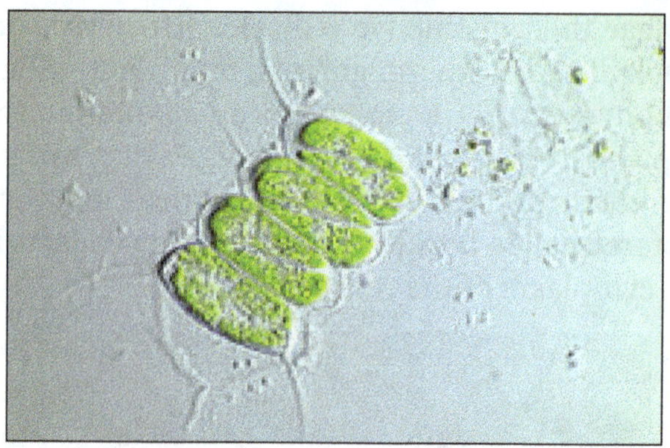

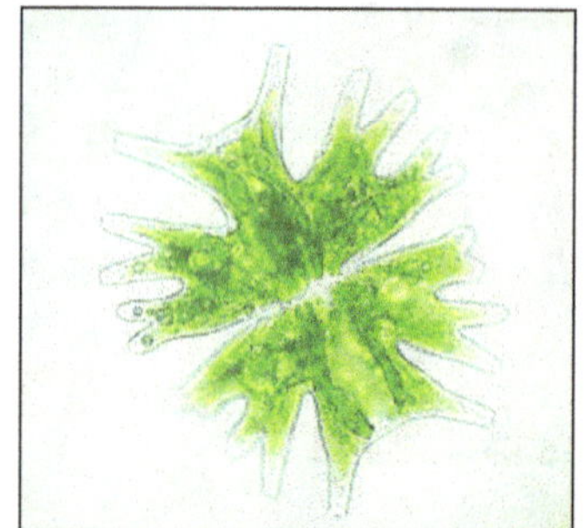

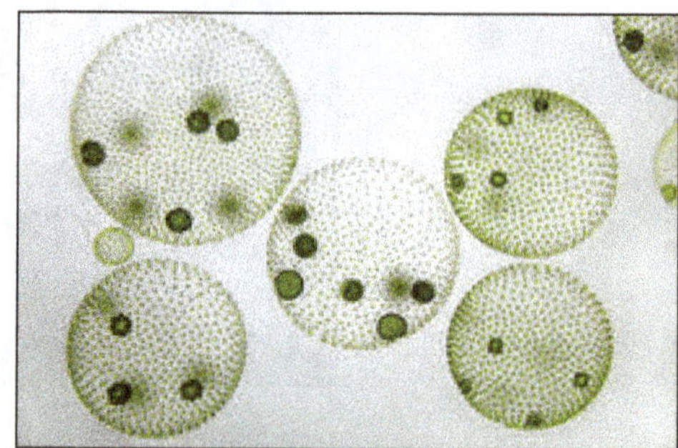

Diversity of green algae. Green algal chloroplasts are characterized by their pigments chlorophyll a and chlorophyll b which give them their green color.

Green algae and plants keep their starch inside their chloroplasts, and in plants and some algae, the chloroplast thylakoids are arranged in grana stacks. Some green algal chloroplasts contain a structure called a pyrenoid, which is functionally similar to the glaucophyte carboxysome in that it is where RuBisCO and CO_2 are concentrated in the chloroplast.

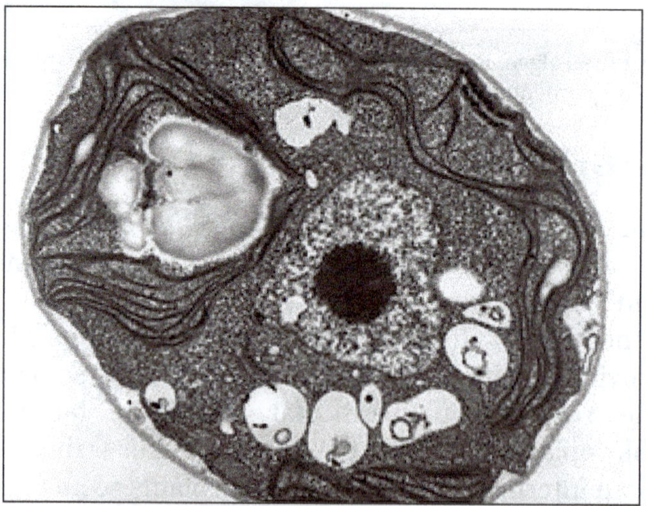

Transmission electron micrograph of Chlamydomonas reinhardtii, a green alga that contains a pyrenoid surrounded by starch.

Helicosporidium is a genus of nonphotosynthetic parasitic green algae that is thought to contain a vestigial chloroplast. Genes from a chloroplast and nuclear genes indicating the presence of a chloroplast have been found in Helicosporidium even if nobody›s seen the chloroplast itself.

Paulinella Chromatophora

While most chloroplasts originate from that first set of endosymbiotic events, Paulinella chromatophora is an exception that acquired a photosynthetic cyanobacterial endosymbiont more recently. It is not clear whether that symbiont is closely related to the ancestral chloroplast of other eukaryotes. Being in the early stages of endosymbiosis, Paulinella chromatophora can offer some insights into how chloroplasts evolved. Paulinella cells contain one or two sausage shaped blue-green photosynthesizing structures called chromatophores, descended from the cyanobacterium Synechococcus. Chromatophores cannot survive outside their host. Chromatophore DNA is about a million base pairs long, containing around 850 protein encoding genes—far less than the three million base pair Synechococcus genome, but much larger than the approximately 150,000 base pair genome of the more assimilated chloroplast. Chromatophores have transferred much less of their DNA to the nucleus of their host. About 0.3–0.8% of the nuclear DNA in Paulinella is from the chromatophore, compared with 11–14% from the chloroplast in plants.

Secondary and Tertiary Endosymbiosis

Many other organisms obtained chloroplasts from the primary chloroplast lineages through secondary endosymbiosis—engulfing a red or green alga that contained a chloroplast. These chloroplasts are known as secondary plastids.

While primary chloroplasts have a double membrane from their cyanobacterial ancestor, secondary chloroplasts have additional membranes outside of the original two, as a result of the secondary endosymbiotic event, when a nonphotosynthetic eukaryote engulfed a chloroplast-containing alga but failed to digest it—much like the cyanobacterium at the beginning of this story. The engulfed alga was broken down, leaving only its chloroplast, and sometimes its cell membrane and nucleus, forming a chloroplast with three or four membranes—the two cyanobacterial membranes, sometimes the eaten alga's cell membrane, and the phagosomal vacuole from the host's cell membrane.

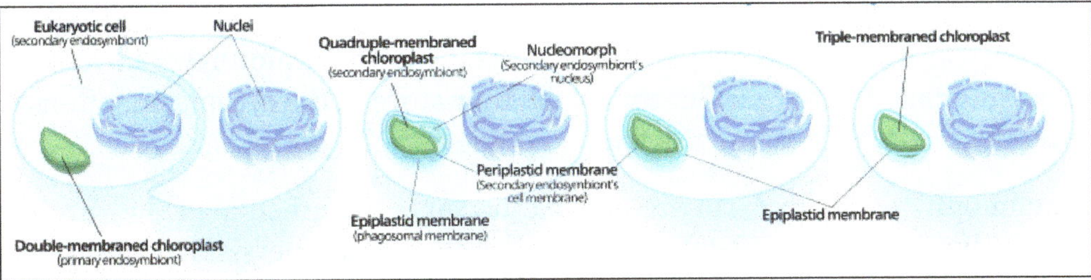

Secondary endosymbiosis consisted of a eukaryotic alga being engulfed by another eukaryote, forming a chloroplast with three or four membranes.

The genes in the phagocytosed eukaryote's nucleus are often transferred to the secondary host's nucleus. Cryptomonads and chlorarachniophytes retain the phagocytosed eukaryote's nucleus, an object called a nucleomorph, located between the second and third membranes of the chloroplast.

All secondary chloroplasts come from green and red algae—no secondary chloroplasts from glaucophytes have been observed, probably because glaucophytes are relatively rare in nature, making them less likely to have been taken up by another eukaryote.

Green Algal Derived Chloroplasts

Green algae have been taken up by the euglenids, chlorarachniophytes, a lineage of dinoflagellates, and possibly the ancestor of the CASH lineage (cryptomonads, alveolates, stramenopiles and haptophytes) in three or four separate engulfments. Many green algal derived chloroplasts contain pyrenoids, but unlike chloroplasts in their green algal ancestors, storage product collects in granules outside the chloroplast.

Euglenophytes

Euglenophytes are a group of common flagellated protists that contain chloroplasts derived from a green alga. Euglenophyte chloroplasts have three membranes—it is thought that the membrane of the primary endosymbiont was lost, leaving the cyanobacterial membranes, and the secondary host's phagosomal membrane. Euglenophyte chloroplasts have a pyrenoid and thylakoids stacked in groups of three. Photosynthetic product is stored in the form of paramylon, which is contained in membrane-bound granules in the cytoplasm of the euglenophyte.

Chlorarachniophytes

Chlorarachniophytes are a rare group of organisms that also contain chloroplasts derived from green algae, though their story is more complicated than that of the euglenophytes. The ancestor of chlorarachniophytes is thought to have been a eukaryote with a red algal derived chloroplast. It is then thought to have lost its first red algal chloroplast, and later engulfed a green alga, giving it its second, green algal derived chloroplast.

Chlorarachniophyte chloroplasts are bounded by four membranes, except near the cell membrane, where the chloroplast membranes fuse into a double membrane. Their thylakoids are arranged in loose stacks of three. Chlorarachniophytes have a form of polysaccharide called chrysolaminarin, which they store in the cytoplasm, often collected around the chloroplast pyrenoid, which bulges into the cytoplasm.

Chlorarachniophyte chloroplasts are notable because the green alga they are derived from has not been completely broken down—its nucleus still persists as a nucleomorph found between the second and third chloroplast membranes—the periplastid space, which corresponds to the green alga's cytoplasm.

Prasinophyte-derived Dinophyte Chloroplast

Lepidodinium viride and its close relatives are dinophytes that lost their original peridinin chloroplast and replaced it with a green algal derived chloroplast (more specifically, a prasinophyte). Lepidodinium is the only dinophyte that has a chloroplast that's not from the rhodoplast lineage. The chloroplast is surrounded by two membranes and has no nucleomorph—all the nucleomorph genes have been transferred to the dinophyte nucleus. The endosymbiotic event that led to this chloroplast was serial secondary endosymbiosis rather than tertiary endosymbiosis—the endosymbiont was a green alga containing a primary chloroplast (making a secondary chloroplast).

Red Algal Derived Chloroplasts

Cryptophytes

Cryptophytes, or cryptomonads are a group of algae that contain a red-algal derived chloroplast. Cryptophyte chloroplasts contain a nucleomorph that superficially resembles that of the chlorarachniophytes. Cryptophyte chloroplasts have four membranes, the outermost of which is continuous with the rough endoplasmic reticulum. They synthesize ordinary starch, which is stored in granules found in the periplastid space—outside the original double membrane, in the place that corresponds to the red alga's cytoplasm. Inside cryptophyte chloroplasts is a pyrenoid and thylakoids in stacks of two.

Their chloroplasts do not have phycobilisomes, but they do have phycobilin pigments which they keep in their thylakoid space, rather than anchored on the outside of their thylakoid membranes.

Cryptophytes may have played a key role in the spreading of red algal based chloroplasts.

Haptophytes

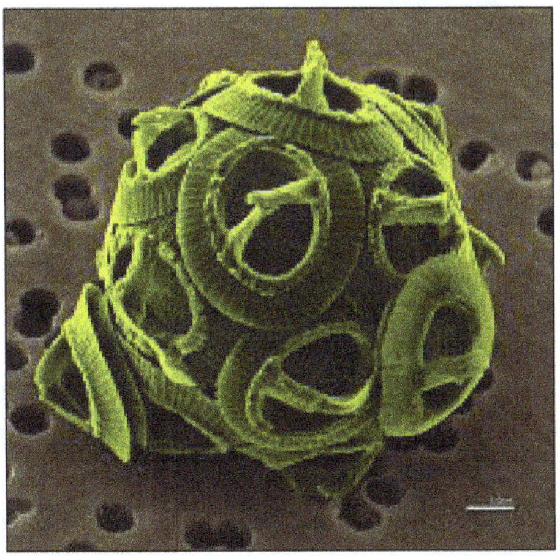

Scanning electron micrograph of Gephyrocapsa oceanica, a haptophyte.

Haptophytes are similar and closely related to cryptophytes or heterokontophytes. Their chloroplasts lack a nucleomorph, their thylakoids are in stacks of three, and they synthesize chrysolaminarin sugar, which they store completely outside of the chloroplast, in the cytoplasm of the haptophyte.

Heterokontophytes (Stramenopiles)

The heterokontophytes, also known as the stramenopiles, are a very large and diverse group of eukaryotes. The photoautotrophic lineage, Ochrophyta, including the diatoms and the brown algae, golden algae, and yellow-green algae, also contains red algal derived chloroplasts.

Heterokont chloroplasts are very similar to haptophyte chloroplasts, containing a pyrenoid, triplet thylakoids, and with some exceptions, having four layer plastidic envelope, the outermost epiplastid membrane connected to the endoplasmic reticulum. Like haptophytes, heterokontophytes store sugar in chrysolaminarin granules in the cytoplasm. Heterokontophyte chloroplasts contain chlorophyll a and with a few exceptions chlorophyll c, but also have carotenoids which give them their many colors.

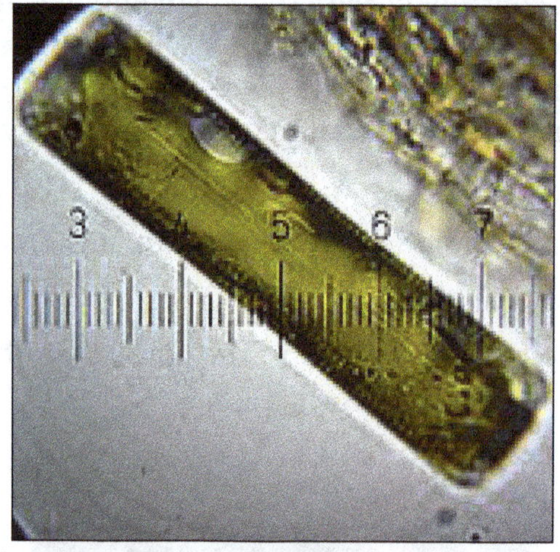

The photosynthetic pigments present in their chloroplasts give diatoms a greenish-brown color.

Apicomplexans, Chromerids and Dinophytes

The alveolates are a major clade of unicellular eukaryotes of both autotrophic and heterotrophic members. The most notable shared characteristic is the presence of cortical (outer-region) alveoli (sacs). These are flattened vesicles (sacs) packed into a continuous layer just under the membrane and supporting it, typically forming a flexible pellicle (thin skin). In dinoflagellates they often form armor plates. Many members contain a red-algal derived plastid. One notable characteristic of this diverse group is the frequent loss of photosynthesis. However, a majority of these heterotrophs continue to process a non-photosynthetic plastid.

Apicomplexans

Apicomplexans are a group of alveolates. Like the helicosproidia, they're parasitic, and have a nonphotosynthetic chloroplast. They were once thought to be related to the helicosproidia, but it is now known that the helicosproida are green algae rather than part of the CASH lineage. The apicomplexans include Plasmodium, the malaria parasite. Many apicomplexans keep a vestigial red algal derived chloroplast called an apicoplast, which they inherited from their ancestors. Other apicomplexans like Cryptosporidium have lost the chloroplast completely. Apicomplexans store their energy in amylopectin granules that are located in their cytoplasm, even though they are non-photosynthetic.

Apicoplasts have lost all photosynthetic function, and contain no photosynthetic pigments or true thylakoids. They are bounded by four membranes, but the membranes are not connected to the endoplasmic reticulum. The fact that apicomplexans still keep their nonphotosynthetic chloroplast around demonstrates how the chloroplast carries out important functions other than photosynthesis. Plant chloroplasts provide plant cells with many important things besides sugar, and apicoplasts are no different—they synthesize fatty acids, isopentenyl pyrophosphate, iron-sulfur clusters, and carry out part of the heme pathway. This makes the apicoplast an attractive target for drugs to cure apicomplexan-related diseases. The most important apicoplast function is isopentenyl pyrophosphate synthesis—in fact, apicomplexans die when something interferes with this apicoplast function, and when apicomplexans are grown in an isopentenyl pyrophosphate-rich medium, they dump the organelle.

Chromerids

The Chromerida is a newly discovered group of algae from Australian corals which comprises some close photosynthetic relatives of the apicomplexans. The first member, Chromera velia, was discovered and first isolated in 2001. The discovery of Chromera velia with similar structure to the apicomplexanss, provides an important link in the evolutionary history of the apicomplexans and dinophytes. Their plastids have four membranes, lack chlorophyll c and use the type II form of RuBisCO obtained from a horizontal transfer event.

Dinophytes

The dinoflagellates are yet another very large and diverse group of protists, around half of which are (at least partially) photosynthetic.

Most dinophyte chloroplasts are secondary red algal derived chloroplasts. Many other dinophytes have lost the chloroplast (becoming the nonphotosynthetic kind of dinoflagellate), or replaced it though tertiary endosymbiosis—the engulfment of another eukaryotic algae containing a red algal derived chloroplast. Others replaced their original chloroplast with a green algal derived one.

Most dinophyte chloroplasts contain form II RuBisCO, at least the photosynthetic pigments chlorophyll a, chlorophyll c_2, beta-carotene, and at least one dinophyte-unique xanthophyll (peridinin, dinoxanthin, or diadinoxanthin), giving many a golden-brown color. All dinophytes store starch in their cytoplasm, and most have chloroplasts with thylakoids arranged in stacks of three.

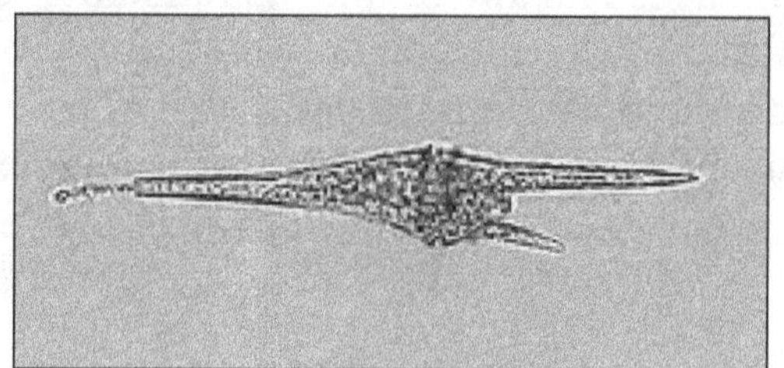

Ceratium furca, a peridinin-containing dinophyte.

The most common dinophyte chloroplast is the peridinin-type chloroplast, characterized by the carotenoid pigment peridinin in their chloroplasts, along with chlorophyll a and chlorophyll c2. Peridinin is not found in any other group of chloroplasts. The peridinin chloroplast is bounded by three membranes (occasionally two), having lost the red algal endosymbiont's original cell membrane. The outermost membrane is not connected to the endoplasmic reticulum. They contain a pyrenoid, and have triplet-stacked thylakoids. Starch is found outside the chloroplast. An important feature of these chloroplasts is that their chloroplast DNA is highly reduced and fragmented into many small circles. Most of the genome has migrated to the nucleus, and only critical photosynthesis-related genes remain in the chloroplast.

The peridinin chloroplast is thought to be the dinophytes' "original" chloroplast, which has been lost, reduced, replaced, or has company in several other dinophyte lineages.

Fucoxanthin-containing (Haptophyte-derived) Dinophyte Chloroplasts

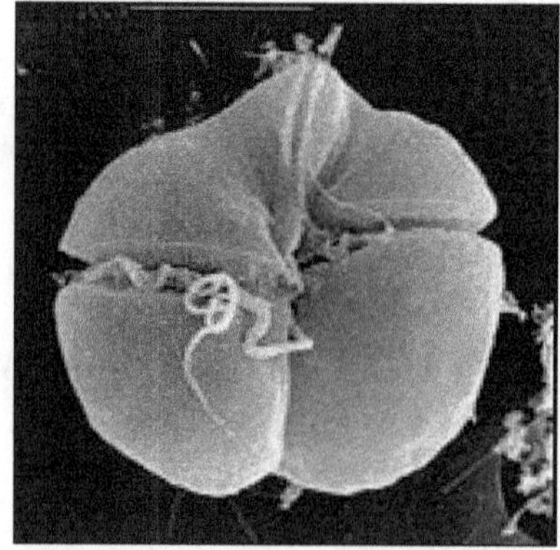

Karenia brevis is a fucoxanthin-containing dynophyte responsible for algal blooms called "red tides".

The fucoxanthin dinophyte lineages (including Karlodinium and Karenia) lost their original red algal derived chloroplast, and replaced it with a new chloroplast derived from a haptophyte endosymbiont. Karlodinium and Karenia probably took up different heterokontophytes. Because the haptophyte chloroplast has four membranes, tertiary endosymbiosis would be expected to create a six membraned chloroplast, adding the haptophyte's cell membrane and the dinophyte's phagosomal vacuole. However, the haptophyte was heavily reduced, stripped of a few membranes and its nucleus, leaving only its chloroplast (with its original double membrane), and possibly one or two additional membranes around it.

Fucoxanthin-containing chloroplasts are characterized by having the pigment fucoxanthin (actually 19′-hexanoyloxy-fucoxanthin and 19′-butanoyloxy-fucoxanthin) and no peridinin. Fucoxanthin is also found in haptophyte chloroplasts, providing evidence of ancestry.

Diatom-derived Dinophyte Chloroplasts

Some dinophytes, like Kryptoperidinium and Durinskia have a diatom (heterokontophyte) derived chloroplast. These chloroplasts are bounded by up to five membranes, (depending on whether you count the entire diatom endosymbiont as the chloroplast, or just the red algal derived chloroplast inside it). The diatom endosymbiont has been reduced relatively little—it still retains its original mitochondria, and has endoplasmic reticulum, ribosomes, a nucleus, and of course, red algal derived chloroplasts—practically a complete cell, all inside the host's endoplasmic reticulum lumen.

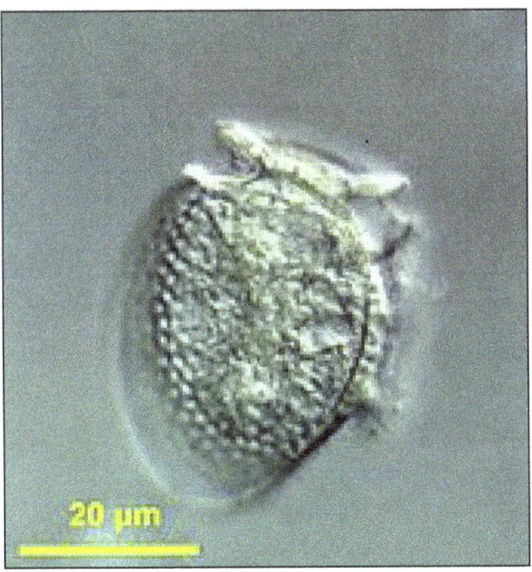

Dinophysis acuminata has chloroplasts taken from a cryptophyte.

However the diatom endosymbiont can't store its own food—its storage polysaccharide is found in granules in the dinophyte host's cytoplasm instead. The diatom endosymbiont's nucleus is present, but it probably can't be called a nucleomorph because it shows no sign of genome reduction, and might have even been expanded. Diatoms have been engulfed by dinoflagellates at least three times.

The diatom endosymbiont is bounded by a single membrane, inside it are chloroplasts with four membranes. Like the diatom endosymbiont's diatom ancestor, the chloroplasts have triplet thylakoids and pyrenoids.

In some of these genera, the diatom endosymbiont's chloroplasts aren't the only chloroplasts in the dinophyte. The original three-membraned peridinin chloroplast is still around, converted to an eyespot.

Kleptoplastidy

In some groups of mixotrophic protists, like some dinoflagellates (e.g. Dinophysis), chloroplasts are separated from a captured alga and used temporarily. These klepto chloroplasts may only have a lifetime of a few days and are then replaced.

Cryptophyte-derived Dinophyte Chloroplast

Members of the genus Dinophysis have a phycobilin-containing chloroplast taken from a cryptophyte. However, the cryptophyte is not an endosymbiont—only the chloroplast seems to have been taken, and the chloroplast has been stripped of its nucleomorph and outermost two membranes, leaving just a two-membraned chloroplast. Cryptophyte chloroplasts require their nucleomorph to maintain themselves, and Dinophysis species grown in cell culture alone cannot survive, so it is possible (but not confirmed) that the Dinophysis chloroplast is a kleptoplast—if so, Dinophysis chloroplasts wear out and Dinophysis species must continually engulf cryptophytes to obtain new chloroplasts to replace the old ones.

Chloroplast DNA

Chloroplasts have their own DNA, often abbreviated as ctDNA, or cpDNA. It is also known as the plastome. Its existence was first proved in 1962, and first sequenced in 1986—when two Japanese research teams sequenced the chloroplast DNA of liverwort and tobacco. Since then, hundreds of chloroplast DNAs from various species have been sequenced, but they are mostly those of land plants and green algae—glaucophytes, red algae, and other algal groups are extremely underrepresented, potentially introducing some bias in views of "typical" chloroplast DNA structure and content.

Molecular Structure

With few exceptions, most chloroplasts have their entire chloroplast genome combined into a single large circular DNA molecule, typically 120,000–170,000 base pairs long. They can have a contour length of around 30–60 micrometers, and have a mass of about 80–130 million daltons.

While usually thought of as a circular molecule, there is some evidence that chloroplast DNA molecules more often take on a linear shape.

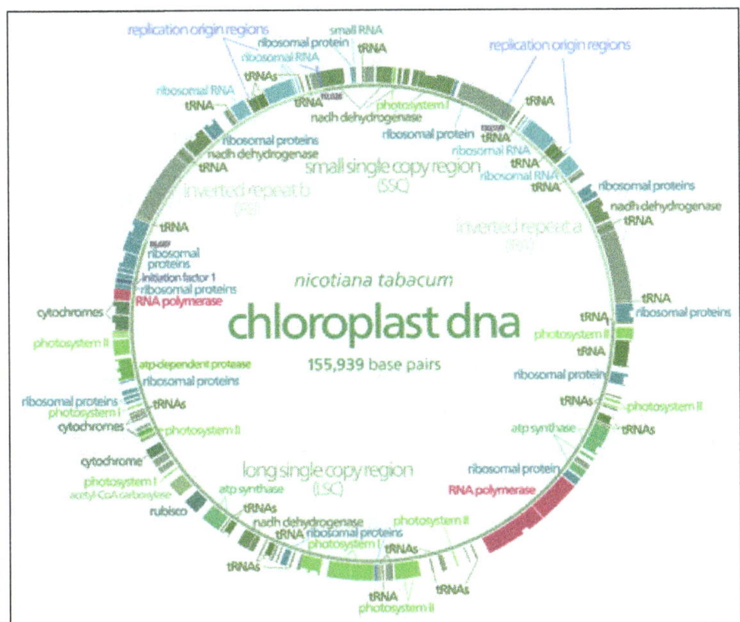

Chloroplast DNA Interactive gene map of chloroplast DNA from Nicotiana tabacum. Segments with labels on the inside reside on the B strand of DNA, segments with labels on the outside are on the A strand. Notches indicate introns.

Inverted Repeats

Many chloroplast DNAs contain two inverted repeats, which separate a long single copy section (LSC) from a short single copy section (SSC). While a given pair of inverted repeats are rarely completely identical, they are always very similar to each other, apparently resulting from concerted evolution.

The inverted repeats vary wildly in length, ranging from 4,000 to 25,000 base pairs long each and containing as few as four or as many as over 150 genes. Inverted repeats in plants tend to be at the upper end of this range, each being 20,000–25,000 base pairs long.

The inverted repeat regions are highly conserved among land plants, and accumulate few mutations. Similar inverted repeats exist in the genomes of cyanobacteria and the other two chloroplast lineages (glaucophyta and rhodophyceae), suggesting that they predate the chloroplast, though some chloroplast DNAs have since lost or flipped the inverted repeats (making them direct repeats). It is possible that the inverted repeats help stabilize the rest of the chloroplast genome, as chloroplast DNAs which have lost some of the inverted repeat segments tend to get rearranged more.

Nucleoids

New chloroplasts may contain up to 100 copies of their DNA, though the number of chloroplast DNA copies decreases to about 15–20 as the chloroplasts age. They are usually packed into nucleoids, which can contain several identical chloroplast DNA rings. Many nucleoids can be found in each chloroplast. In primitive red algae, the chloroplast DNA nucleoids are clustered in the center of the chloroplast, while in green plants and green algae, the nucleoids are dispersed throughout the stroma.

Though chloroplast DNA is not associated with true histones, in red algae, similar proteins that tightly pack each chloroplast DNA ring into a nucleoid have been found.

DNA Repair

In chloroplasts of the moss Physcomitrella patens, the DNA mismatch repair protein Msh1 interacts with the recombinational repair proteins RecA and RecG to maintain chloroplast genome stability. In chloroplasts of the plant Arabidopsis thaliana the RecA protein maintains the integrity of the chloroplast's DNA by a process that likely involves the recombinational repair of DNA damage.

DNA Replication

The Leading Model of cpDNA Replication

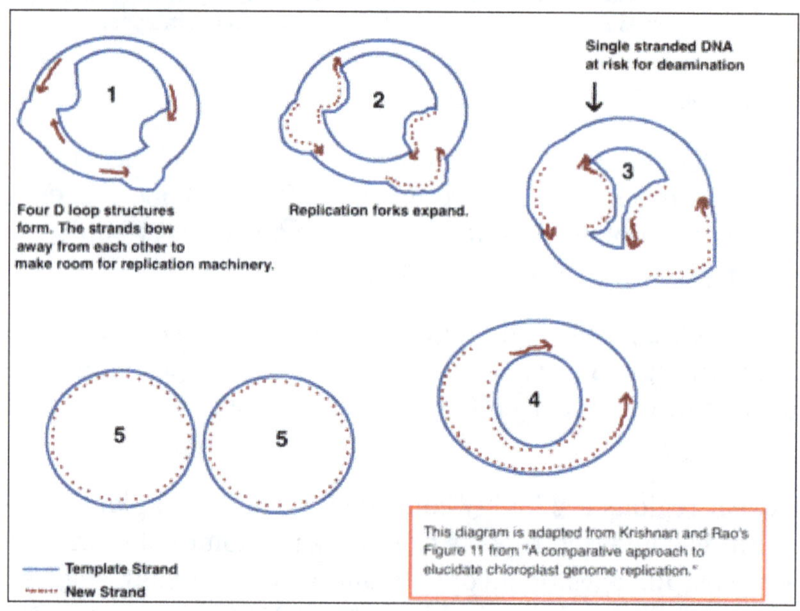

Chloroplast DNA replication via multiple D loop mechanisms.

The mechanism for chloroplast DNA (cpDNA) replication has not been conclusively determined, but two main models have been proposed. Scientists have attempted to observe chloroplast replication via electron microscopy since the 1970s. The results of the microscopy experiments led to the idea that chloroplast DNA replicates using a double displacement loop (D-loop). As the D-loop moves through the circular DNA, it adopts a theta intermediary form, also known as a Cairns replication intermediate, and completes replication with a rolling circle mechanism. Transcription starts at specific points of origin. Multiple replication forks open up, allowing replication machinery to transcribe the DNA. As replication continues, the forks grow and eventually converge. The new cpDNA structures separate, creating daughter cpDNA chromosomes.

In addition to the early microscopy experiments, this model is also supported by the amounts of deamination seen in cpDNA. Deamination occurs when an amino group is lost and is a mutation that often results in base changes. When adenine is deaminated, it becomes hypoxanthine. Hypoxanthine can bind to cytosine, and when the XC base pair is replicated, it becomes a GC (thus, an A → G base change).

Deamination

In cpDNA, there are several A → G deamination gradients. DNA becomes susceptible to deamination events when it is single stranded. When replication forks form, the strand not being copied is single stranded, and thus at risk for A → G deamination. Therefore, gradients in deamination indicate that replication forks were most likely present and the direction that they initially opened (the highest gradient is most likely nearest the start site because it was single stranded for the longest amount of time). This mechanism is still the leading theory today; however, a second theory suggests that most cpDNA is actually linear and replicates through homologous recombination. It further contends that only a minority of the genetic material is kept in circular chromosomes while the rest is in branched, linear, or other complex structures.

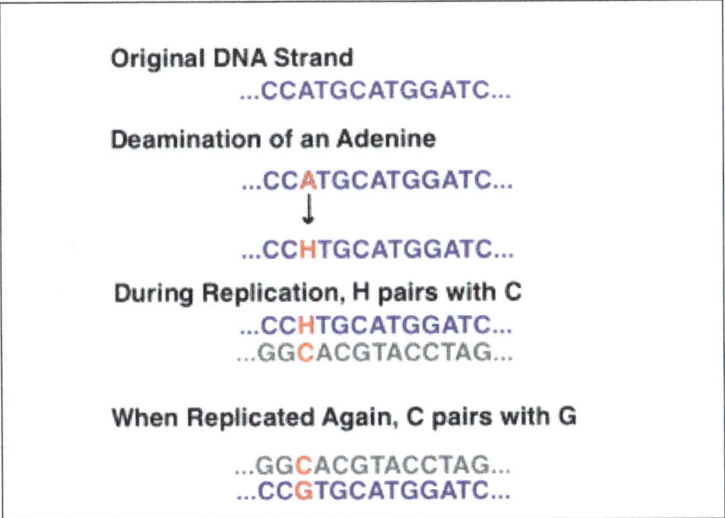

Over time, base changes in the DNA sequence can arise from deamination mutations. When adenine is deaminated, it becomes hypoxanthine, which can pair with cytosine. During replication, the cytosine will pair with guanine, causing an A → G base change.

Alternative Model of Replication

One of competing model for cpDNA replication asserts that most cpDNA is linear and participates in homologous recombination and replication structures similar to the linear and circular DNA structures of bacteriophage T4. It has been established that some plants have linear cpDNA, such as maize, and that more species still contain complex structures that scientists do not yet understand. When the original experiments on cpDNA were performed, scientists did notice linear structures; however, they attributed these linear forms to broken circles. If the branched and complex structures seen in cpDNA experiments are real and not artifacts of concatenated circular DNA or broken circles, then a D-loop mechanism of replication is insufficient to explain how those structures would replicate. At the same time, homologous recombination does not expand the multiple A → G gradients seen in plastomes. Because of the failure to explain the deamination gradient as well as the numerous plant species that have been shown to have circular cpDNA, the predominant theory continues to hold that most cpDNA is circular and most likely replicates via a D loop mechanism.

Gene Content and Protein Synthesis

The chloroplast genome most commonly includes around 100 genes that code for a variety of things, mostly to do with the protein pipeline and photosynthesis. As in prokaryotes, genes in chloroplast DNA are organized into operons. Unlike prokaryotic DNA molecules, chloroplast DNA molecules contain introns (plant mitochondrial DNAs do too, but not human mtDNAs).

Among land plants, the contents of the chloroplast genome are fairly similar.

Chloroplast Genome Reduction and Gene Transfer

Over time, many parts of the chloroplast genome were transferred to the nuclear genome of the host, a process called endosymbiotic gene transfer. As a result, the chloroplast genome is heavily reduced compared to that of free-living cyanobacteria. Chloroplasts may contain 60–100 genes whereas cyanobacteria often have more than 1500 genes in their genome. Recently, a plastid without a genome was found, demonstrating chloroplasts can lose their genome during endosymbiotic the gene transfer process.

Endosymbiotic gene transfer is how we know about the lost chloroplasts in many CASH lineages. Even if a chloroplast is eventually lost, the genes it donated to the former host's nucleus persist, providing evidence for the lost chloroplast's existence. For example, while diatoms (a heterokontophyte) now have a red algal derived chloroplast, the presence of many green algal genes in the diatom nucleus provide evidence that the diatom ancestor had a green algal derived chloroplast at some point, which was subsequently replaced by the red chloroplast.

In land plants, some 11–14% of the DNA in their nuclei can be traced back to the chloroplast, up to 18% in Arabidopsis, corresponding to about 4,500 protein-coding genes. There have been a few recent transfers of genes from the chloroplast DNA to the nuclear genome in land plants.

Of the approximately 3000 proteins found in chloroplasts, some 95% of them are encoded by nuclear genes. Many of the chloroplast's protein complexes consist of subunits from both the chloroplast genome and the host's nuclear genome. As a result, protein synthesis must be coordinated between the chloroplast and the nucleus. The chloroplast is mostly under nuclear control, though chloroplasts can also give out signals regulating gene expression in the nucleus, called retrograde signaling.

Protein Synthesis

Protein synthesis within chloroplasts relies on two RNA polymerases. One is coded by the chloroplast DNA, the other is of nuclear origin. The two RNA polymerases may recognize and bind to different kinds of promoters within the chloroplast genome. The ribosomes in chloroplasts are similar to bacterial ribosomes.

Protein Targeting and Import

Because so many chloroplast genes have been moved to the nucleus, many proteins that would originally have been translated in the chloroplast are now synthesized in the cytoplasm of the plant cell. These proteins must be directed back to the chloroplast, and imported through at least two chloroplast membranes.

Curiously, around half of the protein products of transferred genes aren't even targeted back to the chloroplast. Many became exaptations, taking on new functions like participating in cell division, protein routing, and even disease resistance. A few chloroplast genes found new homes in the mitochondrial genome—most became nonfunctional pseudogenes, though a few tRNA genes still work in the mitochondrion. Some transferred chloroplast DNA protein products get directed to the secretory pathway though many secondary plastids are bounded by an outermost membrane derived from the host's cell membrane, and therefore topologically outside of the cell, because to reach the chloroplast from the cytosol, you have to cross the cell membrane, just like if you were headed for the extracellular space. In those cases, chloroplast-targeted proteins do initially travel along the secretory pathway.

Because the cell acquiring a chloroplast already had mitochondria (and peroxisomes, and a cell membrane for secretion), the new chloroplast host had to develop a unique protein targeting system to avoid having chloroplast proteins being sent to the wrong organelle.

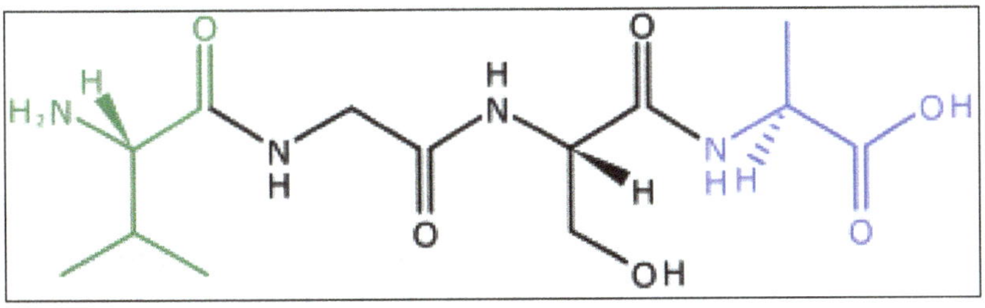

The two ends of a polypeptide are called the N-terminus, or amino end, and the C-terminus, or carboxyl end. This polypeptide has four amino acids linked together. At the left is the N-terminus, with its amino (H_2N) group in green. The blue C-terminus, with its carboxyl group (CO_2H) is at the right.

In most, but not all cases, nuclear-encoded chloroplast proteins are translated with a cleavable transit peptide that's added to the N-terminus of the protein precursor. Sometimes the transit sequence is found on the C-terminus of the protein, or within the functional part of the protein.

Transport Proteins and Membrane Translocons

After a chloroplast polypeptide is synthesized on a ribosome in the cytosol, an enzyme specific to chloroplast proteins phosphorylates, or adds a phosphate group to many (but not all) of them in their transit sequences. Phosphorylation helps many proteins bind the polypeptide, keeping it from folding prematurely. This is important because it prevents chloroplast proteins from assuming their active form and carrying out their chloroplast functions in the wrong place—the cytosol. At the same time, they have to keep just enough shape so that they can be recognized by the chloroplast. These proteins also help the polypeptide get imported into the chloroplast.

From here, chloroplast proteins bound for the stroma must pass through two protein complexes—the TOC complex, or translocon on the outer chloroplast membrane, and the TIC translocon, or translocon on the inner chloroplast membrane translocon. Chloroplast polypeptide chains probably often travel through the two complexes at the same time, but the TIC complex can also retrieve preproteins lost in the intermembrane space.

Structure

In land plants, chloroplasts are generally lens-shaped, 3–10 μm in diameter and 1–3 μm thick. Corn seedling chloroplasts are ≈20 μm3 in volume. Greater diversity in chloroplast shapes exists among the algae, which often contain a single chloroplast that can be shaped like a net (e.g., *Oedogonium*), a cup (e.g., *Chlamydomonas*), a ribbon-like spiral around the edges of the cell (e.g., Spirogyra), or slightly twisted bands at the cell edges (e.g., *Sirogonium*). Some algae have two chloroplasts in each cell; they are star-shaped in *Zygnema*, or may follow the shape of half the cell in order *Desmidiales*. In some algae, the chloroplast takes up most of the cell, with pockets for the nucleus and other organelles, for example, some species of Chlorella have a cup-shaped chloroplast that occupies much of the cell.

All chloroplasts have at least three membrane systems—the outer chloroplast membrane, the inner chloroplast membrane, and the thylakoid system. Chloroplasts that are the product of secondary endosymbiosis may have additional membranes surrounding these three. Inside the outer and inner chloroplast membranes is the chloroplast stroma, a semi-gel-like fluid that makes up much of a chloroplast's volume, and in which the thylakoid system floats.

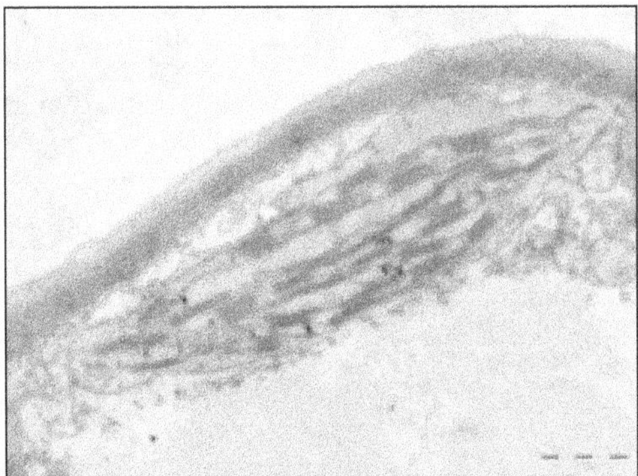

Transmission electron microscope image of a chloroplast. Grana of thylakoids and their connecting lamellae are clearly visible.

There are some common misconceptions about the outer and inner chloroplast membranes. The fact that chloroplasts are surrounded by a double membrane is often cited as evidence that they are the descendants of endosymbiotic cyanobacteria. This is often interpreted as meaning the outer chloroplast membrane is the product of the host's cell membrane infolding to form a vesicle to surround the ancestral cyanobacterium—which is not true—both chloroplast membranes are homologous to the cyanobacterium's original double membranes.

The chloroplast double membrane is also often compared to the mitochondrial double membrane. This is not a valid comparison—the inner mitochondria membrane is used to run proton pumps and carry out oxidative phosphorylation across to generate ATP energy. The only chloroplast structure that can considered analogous to it is the internal thylakoid system. Even so, in terms of "in-out", the direction of chloroplast H+ ion flow is in the opposite direction compared to oxidative phosphorylation in mitochondria. In addition, in terms of function, the inner chloroplast membrane, which regulates metabolite passage and synthesizes some materials, has no counterpart in the mitochondrion.

Outer Chloroplast Membrane

The outer chloroplast membrane is a semi-porous membrane that small molecules and ions can easily diffuse across. However, it is not permeable to larger proteins, so chloroplast polypeptides being synthesized in the cell cytoplasm must be transported across the outer chloroplast membrane by the TOC complex, or translocon on the outer chloroplast membrane.

The chloroplast membranes sometimes protrude out into the cytoplasm, forming a stromule, or stroma-containing tubule. Stromules are very rare in chloroplasts, and are much more common in other plastids like chromoplasts and amyloplasts in petals and roots, respectively. They may exist to increase the chloroplast's surface area for cross-membrane transport, because they are often branched and tangled with the endoplasmic reticulum. When they were first observed in 1962, some plant biologists dismissed the structures as artifactual, claiming that stromules were just oddly shaped chloroplasts with constricted regions or dividing chloroplasts. However, there is a growing body of evidence that stromules are functional, integral features of plant cell plastids, not merely artifacts.

Intermembrane Space and Peptidoglycan Wall

Instead of an intermembrane space, glaucophyte algae have a peptidoglycan wall between their inner and outer chloroplast membranes.

Usually, a thin intermembrane space about 10–20 nanometers thick exists between the outer and inner chloroplast membranes.

Glaucophyte algal chloroplasts have a peptidoglycan layer between the chloroplast membranes. It corresponds to the peptidoglycan cell wall of their cyanobacterial ancestors, which is located between their two cell membranes. These chloroplasts are called muroplasts. Other chloroplasts have lost the cyanobacterial wall, leaving an intermembrane space between the two chloroplast envelope membranes.

Inner Chloroplast Membrane

The inner chloroplast membrane borders the stroma and regulates passage of materials in and out of the chloroplast. After passing through the TOC complex in the outer chloroplast membrane, polypeptides must pass through the TIC complex (translocon on the inner chloroplast membrane) which is located in the inner chloroplast membrane.

In addition to regulating the passage of materials, the inner chloroplast membrane is where fatty acids, lipids, and carotenoids are synthesized.

Peripheral Reticulum

Some chloroplasts contain a structure called the chloroplast peripheral reticulum. It is often found in the chloroplasts of C_4 plants, though it has also been found in some C_3

angiosperms, and even some gymnosperms. The chloroplast peripheral reticulum consists of a maze of membranous tubes and vesicles continuous with the inner chloroplast membrane that extends into the internal stromal fluid of the chloroplast. Its purpose is thought to be to increase the chloroplast's surface area for cross-membrane transport between its stroma and the cell cytoplasm. The small vesicles sometimes observed may serve as transport vesicles to shuttle stuff between the thylakoids and intermembrane space.

Stroma

The protein-rich, alkaline, aqueous fluid within the inner chloroplast membrane and outside of the thylakoid space is called the stroma, which corresponds to the cytosol of the original cyanobacterium. Nucleoids of chloroplast DNA, chloroplast ribosomes, the thylakoid system with plastoglobuli, starch granules, and many proteins can be found floating around in it. The Calvin cycle, which fixes CO_2 into G3P takes place in the stroma.

Chloroplast Ribosomes

Chloroplasts have their own ribosomes, which they use to synthesize a small fraction of their proteins. Chloroplast ribosomes are about two-thirds the size of cytoplasmic ribosomes (around 17 nm vs 25 nm). They take mRNAs transcribed from the chloroplast DNA and translate them into protein. While similar to bacterial ribosomes, chloroplast translation is more complex than in bacteria, so chloroplast ribosomes include some chloroplast-unique features. Small subunit ribosomal RNAs in several Chlorophyta and euglenid chloroplasts lack motifs for shine-dalgarno sequence recognition, which is considered essential for translation initiation in most chloroplasts and prokaryotes. Such loss is also rarely observed in other plastids and prokaryotes.

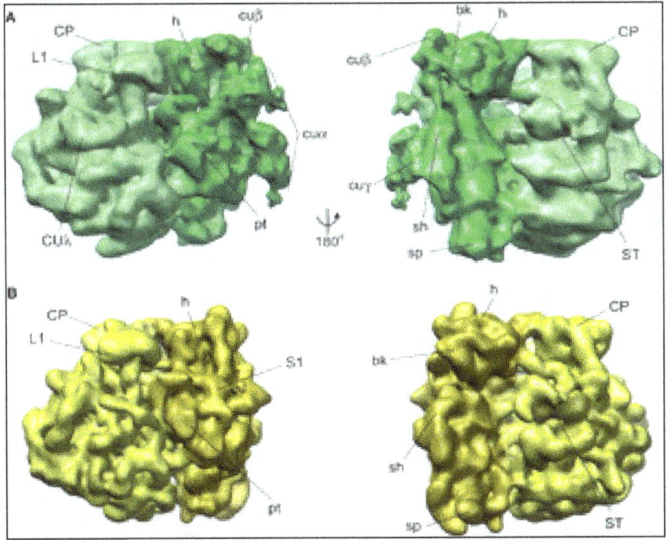

Chloroplast ribosomes Comparison of a chloroplast ribosome (green) and a bacterial ribosome (yellow). Important features common to both ribosomes and chloroplast-unique features are labeled.

Plastoglobuli

Plastoglobuli (singular plastoglobulus, sometimes spelled plastoglobules), are spherical bubbles of lipids and proteins about 45–60 nanometers across. They are surrounded by a lipid monolayer. Plastoglobuli are found in all chloroplasts, but become more common when the chloroplast is under oxidative stress, or when it ages and transitions into a gerontoplast. Plastoglobuli also exhibit a greater size variation under these conditions. They are also common in etioplasts, but decrease in number as the etioplasts mature into chloroplasts.

Plastoglubuli contain both structural proteins and enzymes involved in lipid synthesis and metabolism. They contain many types of lipids including plastoquinone, vitamin E, carotenoids and chlorophylls.

Plastoglobuli were once thought to be free-floating in the stroma, but it is now thought that they are permanently attached either to a thylakoid or to another plastoglobulus attached to a thylakoid, a configuration that allows a plastoglobulus to exchange its contents with the thylakoid network. In normal green chloroplasts, the vast majority of plastoglobuli occur singularly, attached directly to their parent thylakoid. In old or stressed chloroplasts, plastoglobuli tend to occur in linked groups or chains, still always anchored to a thylakoid.

Plastoglobuli form when a bubble appears between the layers of the lipid bilayer of the thylakoid membrane, or bud from existing plastoglubuli—though they never detach and float off into the stroma. Practically all plastoglobuli form on or near the highly curved edges of the thylakoid disks or sheets. They are also more common on stromal thylakoids than on granal ones.

Starch Granules

Starch granules are very common in chloroplasts, typically taking up 15% of the organelle's volume, though in some other plastids like amyloplasts, they can be big enough to distort the shape of the organelle. Starch granules are simply accumulations of starch in the stroma, and are not bounded by a membrane.

Starch granules appear and grow throughout the day, as the chloroplast synthesizes sugars, and are consumed at night to fuel respiration and continue sugar export into the phloem, though in mature chloroplasts, it is rare for a starch granule to be completely consumed or for a new granule to accumulate.

Starch granules vary in composition and location across different chloroplast lineages. In red algae, starch granules are found in the cytoplasm rather than in the chloroplast. In C_4 plants, mesophyll chloroplasts, which do not synthesize sugars, lack starch granules.

RuBisCO

The chloroplast stroma contains many proteins, though the most common and important is RuBisCO, which is probably also the most abundant protein on the planet. RuBisCO is the enzyme that fixes CO_2 into sugar molecules. In C_3 plants, RuBisCO is abundant in all chloroplasts, though in C_4 plants, it is confined to the bundle sheath chloroplasts, where the Calvin cycle is carried out in C_4 plants.

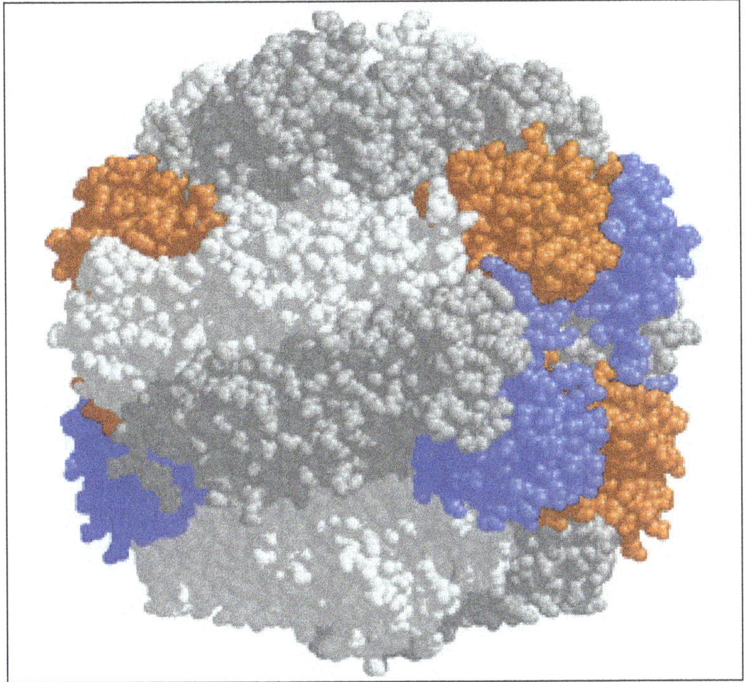

RuBisCO, shown here in a space-filling model, is the main enzyme responsible for carbon fixation in chloroplasts.

Pyrenoids

The chloroplasts of some hornworts and algae contain structures called pyrenoids. They are not found in higher plants. Pyrenoids are roughly spherical and highly refractive bodies which are a site of starch accumulation in plants that contain them. They consist of a matrix opaque to electrons, surrounded by two hemispherical starch plates. The starch is accumulated as the pyrenoids mature. In algae with carbon concentrating mechanisms, the enzyme RuBisCO is found in the pyrenoids. Starch can also accumulate around the pyrenoids when CO_2 is scarce. Pyrenoids can divide to form new pyrenoids, or be produced "de novo".

Thylakoid System

Suspended within the chloroplast stroma is the thylakoid system, a highly dynamic collection of membranous sacks called thylakoids where chlorophyll is found and the light reactions of photosynthesis happen. In most vascular plant chloroplasts, the thylakoids are arranged in stacks called grana, though in certain C_4 plant chloroplasts and some algal chloroplasts, the thylakoids are free floating.

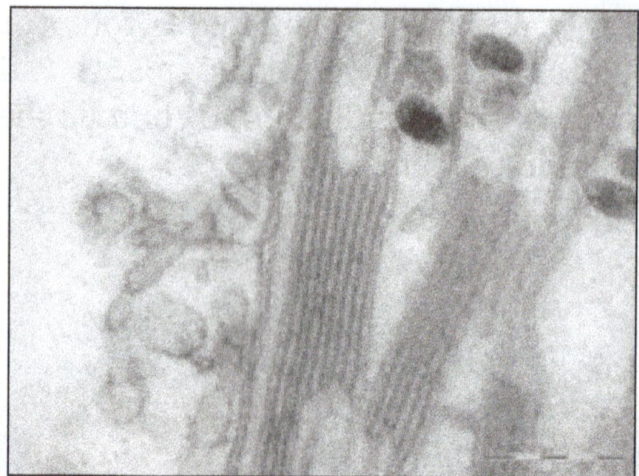

Transmission electron microscope image of some thylakoids arranged in grana stacks and lamellæ. Plastoglobuli (dark blobs) are also present.

Granal Structure

Using a light microscope, it is just barely possible to see tiny green granules—which were named grana. With electron microscopy, it became possible to see the thylakoid system in more detail, revealing it to consist of stacks of flat thylakoids which made up the grana, and long interconnecting stromal thylakoids which linked different grana. In the transmission electron microscope, thylakoid membranes appear as alternating light-and-dark bands, 8.5 nanometers thick.

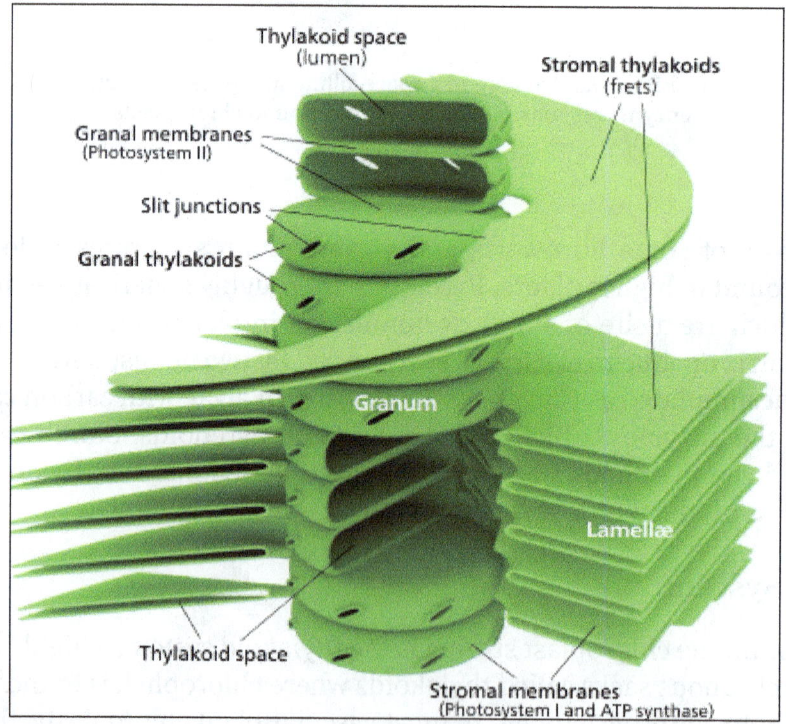

Granum structure The prevailing model for granal structure is a stack of granal thylakoids linked by helical stromal thylakoids that wrap around the grana stacks and form large sheets that connect different grana.

For a long time, the three-dimensional structure of the thylakoid system has been unknown or disputed. One model has the granum as a stack of thylakoids linked by helical stromal thylakoids; the other has the granum as a single folded thylakoid connected in a "hub and spoke" way to other grana by stromal thylakoids. While the thylakoid system is still commonly depicted according to the folded thylakoid model, it was determined in 2011 that the stacked and helical thylakoids model is correct.

In the helical thylakoid model, grana consist of a stack of flattened circular granal thylakoids that resemble pancakes. Each granum can contain anywhere from two to a hundred thylakoids, though grana with 10–20 thylakoids are most common. Wrapped around the grana are helicoid stromal thylakoids, also known as frets or lamellar thylakoids. The helices ascend at an angle of 20–25°, connecting to each granal thylakoid at a bridge-like slit junction. The helicoids may extend as large sheets that link multiple grana, or narrow to tube-like bridges between grana. While different parts of the thylakoid system contain different membrane proteins, the thylakoid membranes are continuous and the thylakoid space they enclose form a single continuous labyrinth.

Thylakoids

Thylakoids are small interconnected sacks which contain the membranes that the light reactions of photosynthesis take place on. Embedded in the thylakoid membranes are important protein complexes which carry out the light reactions of photosynthesis. Photosystem II and photosystem I contain light-harvesting complexes with chlorophyll and carotenoids that absorb light energy and use it to energize electrons. Molecules in the thylakoid membrane use the energized electrons to pump hydrogen ions into the thylakoid space, decreasing the pH and turning it acidic. ATP synthase is a large protein complex that harnesses the concentration gradient of the hydrogen ions in the thylakoid space to generate ATP energy as the hydrogen ions flow back out into the stroma—much like a dam turbine.

There are two types of thylakoids—granal thylakoids, which are arranged in grana, and stromal thylakoids, which are in contact with the stroma. Granal thylakoids are pancake-shaped circular disks about 300–600 nanometers in diameter. Stromal thylakoids are helicoid sheets that spiral around grana. The flat tops and bottoms of granal thylakoids contain only the relatively flat photosystem II protein complex. This allows them to stack tightly, forming grana with many layers of tightly appressed membrane, called granal membrane, increasing stability and surface area for light capture.

In contrast, photosystem I and ATP synthase are large protein complexes which jut out into the stroma. They can't fit in the appressed granal membranes, and so are found in the stromal thylakoid membrane—the edges of the granal thylakoid disks and the stromal thylakoids. These large protein complexes may act as spacers between the sheets of stromal thylakoids.

The number of thylakoids and the total thylakoid area of a chloroplast is influenced by light exposure. Shaded chloroplasts contain larger and more grana with more thylakoid membrane area than chloroplasts exposed to bright light, which have smaller and fewer grana and less thylakoid area. Thylakoid extent can change within minutes of light exposure or removal.

Pigments and Chloroplast Colors

Inside the photosystems embedded in chloroplast thylakoid membranes are various photosynthetic pigments, which absorb and transfer light energy. The types of pigments found are different in various groups of chloroplasts, and are responsible for a wide variety of chloroplast colorations.

Chlorophylls

Chlorophyll a is found in all chloroplasts, as well as their cyanobacterial ancestors. Chlorophyll a is a blue-green pigment partially responsible for giving most cyanobacteria and chloroplasts their color. Other forms of chlorophyll exist, such as the accessory pigments chlorophyll b, chlorophyll c, chlorophyll d, and chlorophyll f.

Chlorophyll b is an olive green pigment found only in the chloroplasts of plants, green algae, any secondary chloroplasts obtained through the secondary endosymbiosis of a green alga, and a few cyanobacteria. It is the chlorophylls a and b together that make most plant and green algal chloroplasts green.

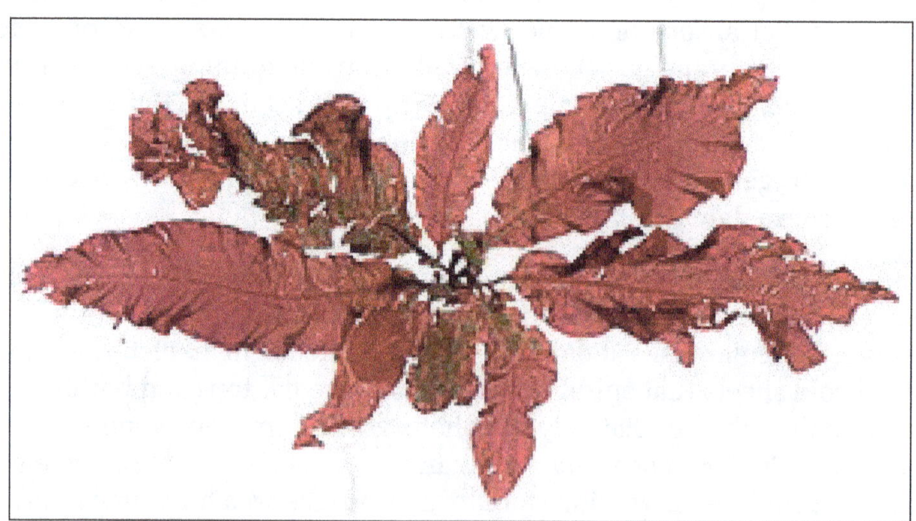

Delesseria sanguinea, a red alga, has chloroplasts that contain red pigments like phycoerytherin that mask their blue-green chlorophyll a.

Chlorophyll c is mainly found in secondary endosymbiotic chloroplasts that originated from a red alga, although it is not found in chloroplasts of red algae themselves. Chlorophyll c is also found in some green algae and cyanobacteria.

Chlorophylls d and f are pigments found only in some cyanobacteria.

Carotenoids

In addition to chlorophylls, another group of yellow–orange pigments called carotenoids are also found in the photosystems. There are about thirty photosynthetic carotenoids. They help transfer and dissipate excess energy, and their bright colors sometimes override the chlorophyll green, like during the fall, when the leaves of some land plants change color. β-carotene is a bright red-orange carotenoid found in nearly all chloroplasts, like chlorophyll a. Xanthophylls, especially the orange-red zeaxanthin, are also common. Many other forms of carotenoids exist that are only found in certain groups of chloroplasts.

Phycobilins

Phycobilins are a third group of pigments found in cyanobacteria, and glaucophyte, red algal, and cryptophyte chloroplasts. Phycobilins come in all colors, though phycoerytherin is one of the pigments that makes many red algae red. Phycobilins often organize into relatively large protein complexes about 40 nanometers across called phycobilisomes. Like photosystem I and ATP synthase, phycobilisomes jut into the stroma, preventing thylakoid stacking in red algal chloroplasts. Cryptophyte chloroplasts and some cyanobacteria don't have their phycobilin pigments organized into phycobilisomes, and keep them in their thylakoid space instead.

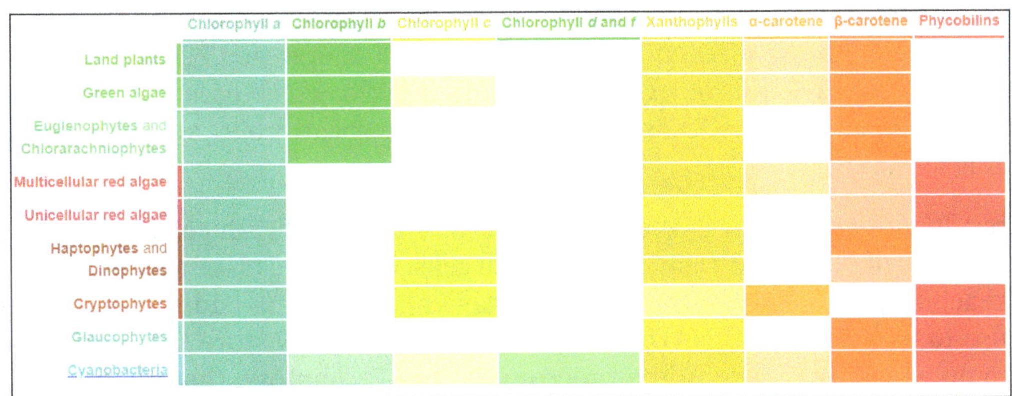

Photosynthetic pigments: Table of the presence of various pigments across chloroplast groups. Colored cells represent pigment presence.

Specialized Chloroplasts in C$_4$ Plants

To fix carbon dioxide into sugar molecules in the process of photosynthesis, chloroplasts use an enzyme called RuBisCO. RuBisCO has a problem—it has trouble distinguishing between carbon dioxide and oxygen, so at high oxygen concentrations, RuBisCO starts accidentally adding oxygen to sugar precursors. This has the end result of ATP energy being wasted and CO_2 being released, all with no sugar being produced. This is a big problem, since O_2 is produced by the initial light reactions of photosynthesis, causing issues down the line in the Calvin cycle which uses RuBisCO.

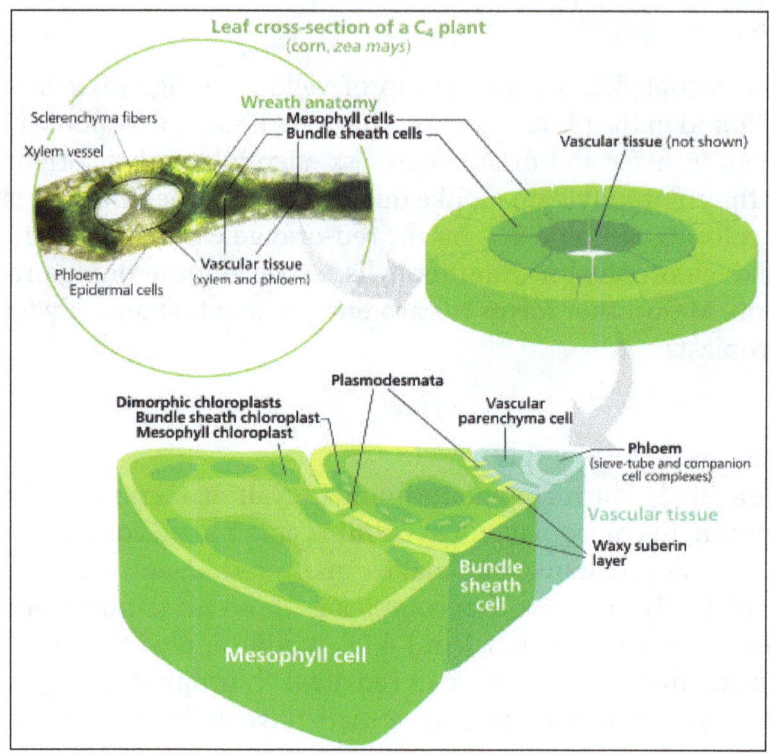

C_4 plants evolved a way to solve this—by spatially separating the light reactions and the Calvin cycle. The light reactions, which store light energy in ATP and NADPH, are done in the mesophyll cells of a C_4 leaf. The Calvin cycle, which uses the stored energy to make sugar using RuBisCO, is done in the bundle sheath cells, a layer of cells surrounding a vein in a leaf.

As a result, chloroplasts in C_4 mesophyll cells and bundle sheath cells are specialized for each stage of photosynthesis. In mesophyll cells, chloroplasts are specialized for the light reactions, so they lack RuBisCO, and have normal grana and thylakoids, which they use to make ATP and NADPH, as well as oxygen. They store CO_2 in a four-carbon compound, which is why the process is called C_4 photosynthesis. The four-carbon compound is then transported to the bundle sheath chloroplasts, where it drops off CO_2 and returns to the mesophyll. Bundle sheath chloroplasts do not carry out the light reactions, preventing oxygen from building up in them and disrupting RuBisCO activity. Because of this, they lack thylakoids organized into grana stacks—though bundle sheath chloroplasts still have free-floating thylakoids in the stroma where they still carry out cyclic electron flow, a light-driven method of synthesizing ATP to power the Calvin cycle without generating oxygen. They lack photosystem II, and only have photosystem I—the only protein complex needed for cyclic electron flow. Because the job of bundle sheath chloroplasts is to carry out the Calvin cycle and make sugar, they often contain large starch grains.

Both types of chloroplast contain large amounts of chloroplast peripheral reticulum, which they use to get more surface area to transport stuff in and out of them. Mesophyll chloroplasts have a little more peripheral reticulum than bundle sheath chloroplasts.

Location

Distribution in a Plant

Not all cells in a multicellular plant contain chloroplasts. All green parts of a plant contain chloroplasts—the chloroplasts, or more specifically, the chlorophyll in them are what make the photosynthetic parts of a plant green. The plant cells which contain chloroplasts are usually parenchyma cells, though chloroplasts can also be found in collenchyma tissue. A plant cell which contains chloroplasts is known as a chlorenchyma cell. A typical chlorenchyma cell of a land plant contains about 10 to 100 chloroplasts.

In some plants such as cacti, chloroplasts are found in the stems, though in most plants, chloroplasts are concentrated in the leaves. One square millimeter of leaf tissue can contain half a million chloroplasts. Within a leaf, chloroplasts are mainly found in the mesophyll layers of a leaf, and the guard cells of stomata. Palisade mesophyll cells can contain 30–70 chloroplasts per cell, while stomatal guard cells contain only around 8–15 per cell, as well as much less chlorophyll. Chloroplasts can also be found in the bundle sheath cells of a leaf, especially in C_4 plants, which carry out the Calvin cycle in their bundle sheath cells. They are often absent from the epidermis of a leaf.

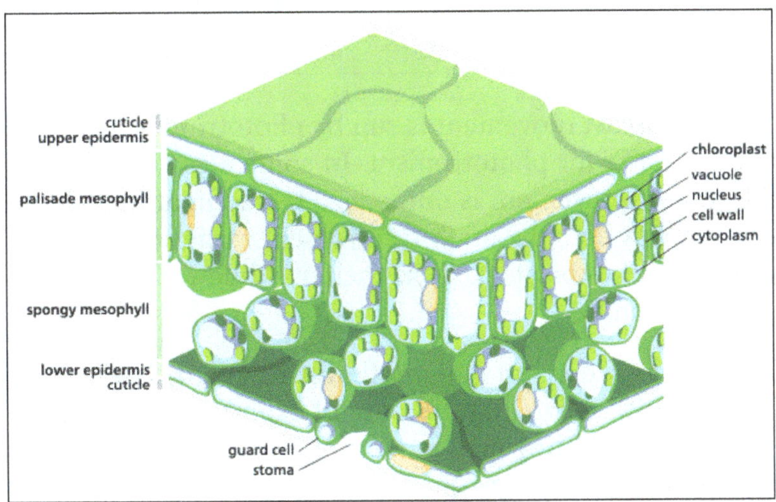

A cross section of a leaf, showing chloroplasts in its mesophyll cells. Stomal guard cells also have chloroplasts, though much fewer than mesophyll cells.

Cellular Location

Chloroplast Movement

The chloroplasts of plant and algal cells can orient themselves to best suit the available light. In low-light conditions, they will spread out in a sheet—maximizing the surface area to absorb light. Under intense light, they will seek shelter by aligning in vertical columns along the plant cell's cell wall or turning sideways so that light strikes them edge-on. This reduces exposure and protects them from photooxidative damage. This ability to distribute chloroplasts so that they can take shelter behind each other or spread out may be the reason why land plants evolved to have many small chloroplasts instead of a few big ones. Chloroplast movement is considered one of the most closely regulated stimulus-response systems that can be found in plants. Mitochondria have also been observed to follow chloroplasts as they move.

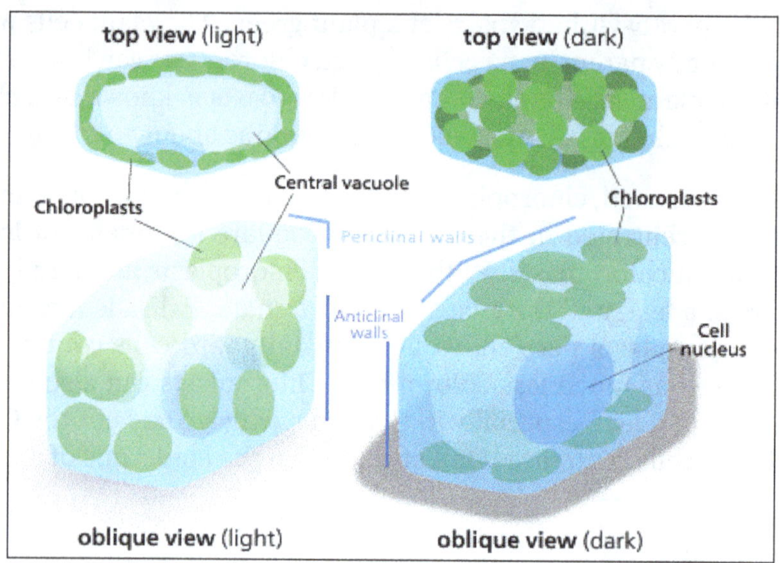

In higher plants, chloroplast movement is run by phototropins, blue light photoreceptors also responsible for plant phototropism. In some algae, mosses, ferns, and flowering plants, chloroplast movement is influenced by red light in addition to blue light, though very long red wavelengths inhibit movement rather than speeding it up. Blue light generally causes chloroplasts to seek shelter, while red light draws them out to maximize light absorption.

Studies of Vallisneria gigantea, an aquatic flowering plant, have shown that chloroplasts can get moving within five minutes of light exposure, though they don't initially show any net directionality. They may move along microfilament tracks, and the fact that the microfilament mesh changes shape to form a honeycomb structure surrounding the chloroplasts after they have moved suggests that microfilaments may help to anchor chloroplasts in place.

Function and Chemistry

Guard Cell Chloroplasts

Unlike most epidermal cells, the guard cells of plant stomata contain relatively well-developed chloroplasts. However, exactly what they do is controversial.

Plant Innate Immunity

Plants lack specialized immune cells—all plant cells participate in the plant immune response. Chloroplasts, along with the nucleus, cell membrane, and endoplasmic reticulum, are key players in pathogen defense. Due to its role in a plant cell's immune response, pathogens frequently target the chloroplast.

Plants have two main immune responses—the hypersensitive response, in which infected cells seal themselves off and undergo programmed cell death, and systemic acquired resistance, where infected cells release signals warning the rest of the plant of a pathogen's presence. Chloroplasts stimulate both responses by purposely damaging their photosynthetic system, producing reactive oxygen species. High levels of reactive oxygen species will cause the hypersensitive response. The reactive oxygen species also directly kill any pathogens within the cell. Lower levels of reactive oxygen species initiate systemic acquired resistance, triggering defense-molecule production in the rest of the plant.

In some plants, chloroplasts are known to move closer to the infection site and the nucleus during an infection.

Chloroplasts can serve as cellular sensors. After detecting stress in a cell, which might be due to a pathogen, chloroplasts begin producing molecules like salicylic acid, jasmonic acid, nitric oxide and reactive oxygen species which can serve as defense-signals. As cellular signals, reactive oxygen species are unstable molecules, so they probably don't leave the chloroplast, but instead pass on their signal to an unknown second messenger molecule. All these molecules initiate retrograde signaling—signals from the chloroplast that regulate gene expression in the nucleus.

In addition to defense signaling, chloroplasts, with the help of the peroxisomes, help synthesize an important defense molecule, jasmonate. Chloroplasts synthesize all the fatty acids in a plant cell—linoleic acid, a fatty acid, is a precursor to jasmonate.

Photosynthesis

One of the main functions of the chloroplast is its role in photosynthesis, the process by which light is transformed into chemical energy, to subsequently produce food in the form of sugars. Water (H_2O) and carbon dioxide (CO_2) are used in photosynthesis, and sugar and oxygen (O_2) is made, using light energy. Photosynthesis is divided into two stages—the light reactions, where water is split to produce oxygen, and the dark reactions, or Calvin cycle, which builds sugar molecules from carbon dioxide. The two phases are linked by the energy carriers adenosine triphosphate (ATP) and nicotinamide adenine dinucleotide phosphate ($NADP^+$).

Light Reactions

The light reactions take place on the thylakoid membranes. They take light energy and store it in NADPH, a form of NADP+, and ATP to fuel the dark reactions.

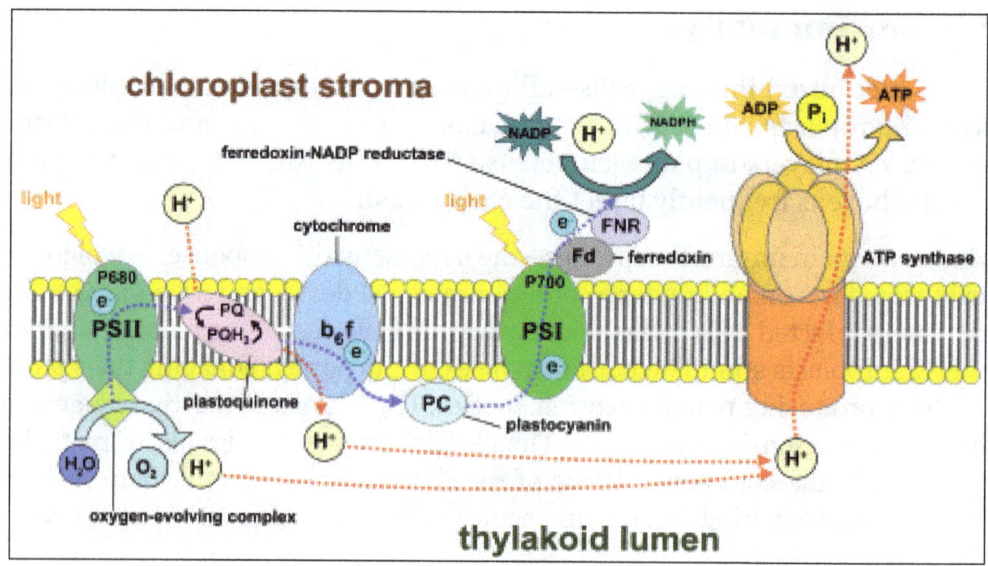

The light reactions of photosynthesis take place across the thylakoid membranes.

Energy Carriers

ATP is the phosphorylated version of adenosine diphosphate (ADP), which stores energy in a cell and powers most cellular activities. ATP is the energized form, while ADP is the (partially) depleted form. $NADP^+$ is an electron carrier which ferries high energy electrons. In the light reactions, it gets reduced, meaning it picks up electrons, becoming $NADPH^+$.

Photophosphorylation

Like mitochondria, chloroplasts use the potential energy stored in an H^+, or hydrogen ion gradient to generate ATP energy. The two photosystems capture light energy to energize electrons taken from water, and release them down an electron transport chain. The molecules between the photosystems harness the electrons' energy to pump hydrogen ions into the thylakoid space, creating a concentration gradient, with more hydrogen ions (up to a thousand times as many) inside the thylakoid system than in the stroma. The hydrogen ions in the thylakoid space then diffuse back down their concentration gradient, flowing back out into the stroma through ATP synthase. ATP synthase uses the energy from the flowing hydrogen ions to phosphorylate adenosine diphosphate into adenosine triphosphate, or ATP. Because chloroplast ATP synthase projects out into the stroma, the ATP is synthesized there, in position to be used in the dark reactions.

$NADP^+$ Reduction

Electrons are often removed from the electron transport chains to charge $NADP^+$ with electrons, reducing it to NADPH. Like ATP synthase, ferredoxin-NADP+ reductase, the enzyme that reduces NADP$^+$, releases the NADPH it makes into the stroma, right where it is needed for the dark reactions.

Because NADP$^+$ reduction removes electrons from the electron transport chains, they must be replaced—the job of photosystem II, which splits water molecules (H_2O) to obtain the electrons from its hydrogen atoms.

Cyclic Photophosphorylation

While photosystem II photolyzes water to obtain and energize new electrons, photosystem I simply reenergizes depleted electrons at the end of an electron transport chain. Normally, the reenergized electrons are taken by NADP$^+$, though sometimes they can flow back down more H$^+$-pumping electron transport chains to transport more hydrogen ions into the thylakoid space to generate more ATP. This is termed cyclic photophosphorylation because the electrons are recycled. Cyclic photophosphorylation is common in C$_4$ plants, which need more ATP than NADPH.

Dark Reactions

The Calvin cycle, also known as the dark reactions, is a series of biochemical reactions that fixes CO_2 into G3P sugar molecules and uses the energy and electrons from the ATP and NADPH made in the light reactions. The Calvin cycle takes place in the stroma of the chloroplast.

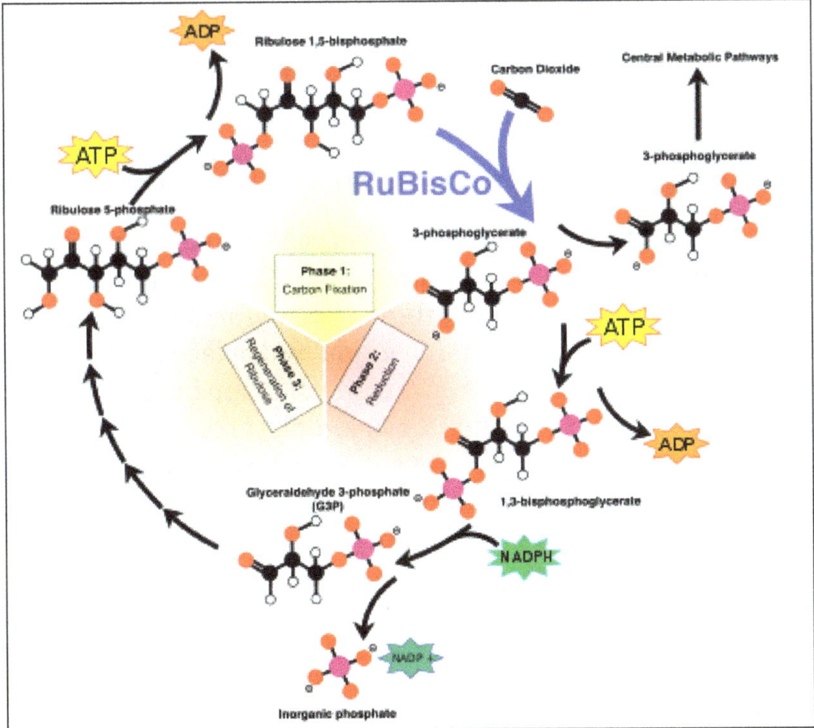

The Calvin cycle: The Calvin cycle incorporates carbon dioxide into sugar molecules.

While named "the dark reactions", in most plants, they take place in the light, since the dark reactions are dependent on the products of the light reactions.

Carbon Fixation and G3P Synthesis

The Calvin cycle starts by using the enzyme RuBisCO to fix CO_2 into five-carbon Ribulose bisphosphate (RuBP) molecules. The result is unstable six-carbon molecules that immediately break down into three-carbon molecules called 3-phosphoglyceric acid, or 3-PGA. The ATP and NADPH made in the light reactions is used to convert the 3-PGA into glyceraldehyde-3-phosphate, or G3P sugar molecules. Most of the G3P molecules are recycled back into RuBP using energy from more ATP, but one out of every six produced leaves the cycle—the end product of the dark reactions.

Sugars and Starches

Glyceraldehyde-3-phosphate can double up to form larger sugar molecules like glucose and fructose. These molecules are processed, and from them, the still larger sucrose, a disaccharide commonly known as table sugar, is made, though this process takes place outside of the chloroplast, in the cytoplasm.

Sucrose is made up of a glucose monomer (left), and a fructose monomer (right).

Alternatively, glucose monomers in the chloroplast can be linked together to make starch, which accumulates into the starch grains found in the chloroplast. Under conditions such as high atmospheric CO_2 concentrations, these starch grains may grow very large, distorting the grana and thylakoids. The starch granules displace the thylakoids, but leave them intact. Waterlogged roots can also cause starch buildup in the chloroplasts, possibly due to less sucrose being exported out of the chloroplast (or more accurately, the plant cell). This depletes a plant's free phosphate supply, which indirectly stimulates chloroplast starch synthesis. While linked to low photosynthesis rates, the starch grains themselves may not necessarily interfere significantly with the efficiency of photosynthesis, and might simply be a side effect of another photosynthesis-depressing factor.

Photorespiration

Photorespiration can occur when the oxygen concentration is too high. RuBisCO cannot distinguish between oxygen and carbon dioxide very well, so it can accidentally add

O_2 instead of CO_2 to RuBP. This process reduces the efficiency of photosynthesis—it consumes ATP and oxygen, releases CO_2, and produces no sugar. It can waste up to half the carbon fixed by the Calvin cycle. Several mechanisms have evolved in different lineages that raise the carbon dioxide concentration relative to oxygen within the chloroplast, increasing the efficiency of photosynthesis. These mechanisms are called carbon dioxide concentrating mechanisms, or CCMs. These include Crassulacean acid metabolism, C_4 carbon fixation, and pyrenoids. Chloroplasts in C4 plants are notable as they exhibit a distinct chloroplast dimorphism.

pH

Because of the H+ gradient across the thylakoid membrane, the interior of the thylakoid is acidic, with a pH around 4, while the stroma is slightly basic, with a pH of around 8. The optimal stroma pH for the Calvin cycle is 8.1, with the reaction nearly stopping when the pH falls below 7.3.

CO_2 in water can form carbonic acid, which can disturb the pH of isolated chloroplasts, interfering with photosynthesis, even though CO_2 is used in photosynthesis. However, chloroplasts in living plant cells are not affected by this as much.

Chloroplasts can pump K^+ and H^+ ions in and out of themselves using a poorly understood light-driven transport system.

In the presence of light, the pH of the thylakoid lumen can drop up to 1.5 pH units, while the pH of the stroma can rise by nearly one pH unit.

Amino Acid Synthesis

Chloroplasts alone make almost all of a plant cell's amino acids in their stroma except the sulfur-containing ones like cysteine and methionine. Cysteine is made in the chloroplast (the proplastid too) but it is also synthesized in the cytosol and mitochondria, probably because it has trouble crossing membranes to get to where it is needed. The chloroplast is known to make the precursors to methionine but it is unclear whether the organelle carries out the last leg of the pathway or if it happens in the cytosol.

Other Nitrogen Compounds

Chloroplasts make all of a cell's purines and pyrimidines—the nitrogenous bases found in DNA and RNA. They also convert nitrite (NO_2-) into ammonia (NH_3) which supplies the plant with nitrogen to make its amino acids and nucleotides.

Other Chemical Products

The plastid is the site of diverse and complex lipid synthesis in plants. The carbon used to form the majority of the lipid is from acetyl-CoA, which is the decarboxylation product of pyruvate. Pyruvate may enter the plastid from the cytosol by passive diffusion through the membrane after production in glycolysis. Pyruvate is also made in the plastid from phosphoenolpyruvate, a metabolite made in the cytosol from pyruvate or PGA. Acetate in the cytosol is unavailable for lipid biosynthesis in the plastid. The typical length of fatty acids produced in the plastid are 16 or 18 carbons, with 0-3 cis double bonds.

The biosynthesis of fatty acids from acetyl-CoA primarily requires two enzymes. Acetyl-CoA carboxylase creates malonyl-CoA, used in both the first step and the extension steps of synthesis. Fatty acid synthase (FAS) is a large complex of enzymes and co-factors including acyl carrier protein (ACP) which holds the acyl chain as it is synthesized. The initiation of synthesis begins with the condensation of malonyl-ACP with acetyl-CoA to produce ketobutyryl-ACP. 2 reductions involving the use of NADPH and one dehydration creates butyryl-ACP. Extension of the fatty acid comes from repeated cycles of malonyl-ACP condensation, reduction, and dehydration.

Other lipids are derived from the methyl-erythritol phosphate (MEP) pathway and consist of gibberelins, sterols, abscisic acid, phytol, and innumerable secondary metabolites.

Differentiation, Replication and Inheritance

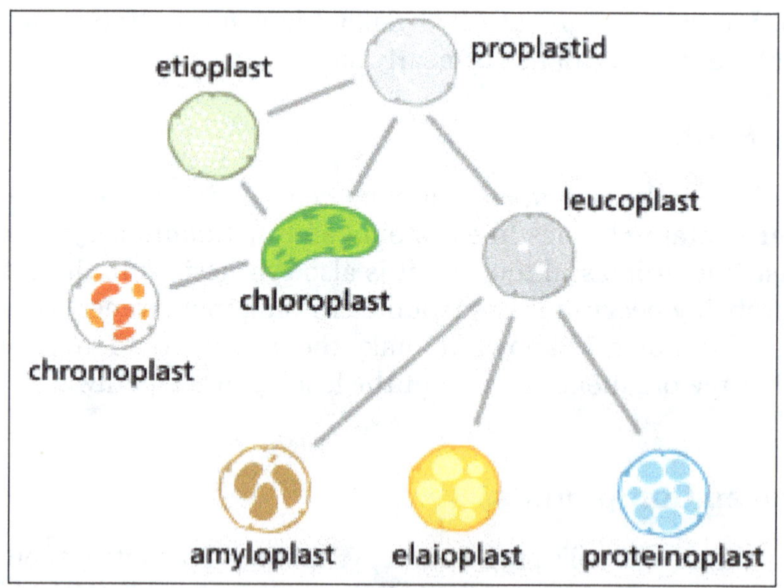

Plastid types: Plants contain many different kinds of plastids in their cells.

Chloroplasts are a special type of a plant cell organelle called a plastid, though the two terms are sometimes used interchangeably. There are many other types of plastids, which carry out various functions. All chloroplasts in a plant are descended from undifferentiated proplastids found in the zygote, or fertilized egg. Proplastids are commonly found in an adult plant's apical meristems. Chloroplasts do not normally develop from proplastids in root tip meristems—instead, the formation of starch-storing amyloplasts is more common.

In shoots, proplastids from shoot apical meristems can gradually develop into chloroplasts in photosynthetic leaf tissues as the leaf matures, if exposed to the required light. This process involves invaginations of the inner plastid membrane, forming sheets of membrane that project into the internal stroma. These membrane sheets then fold to form thylakoids and grana.

If angiosperm shoots are not exposed to the required light for chloroplast formation, proplastids may develop into an etioplast stage before becoming chloroplasts. An etioplast is a plastid that lacks chlorophyll, and has inner membrane invaginations that form a lattice of tubes in their stroma, called a prolamellar body. While etioplasts lack chlorophyll, they have a yellow chlorophyll precursor stocked. Within a few minutes of light exposure, the prolamellar body begins to reorganize into stacks of thylakoids, and chlorophyll starts to be produced. This process, where the etioplast becomes a chloroplast, takes several hours. Gymnosperms do not require light to form chloroplasts.

Light, however, does not guarantee that a proplastid will develop into a chloroplast. Whether a proplastid develops into a chloroplast some other kind of plastid is mostly controlled by the nucleus and is largely influenced by the kind of cell it resides in.

Plastid Interconversion

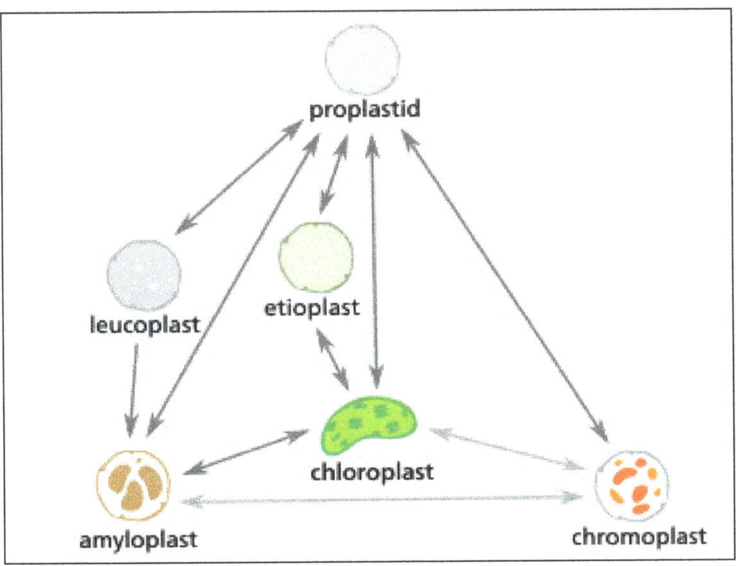

Many plastid interconversions are possible.

Plastid differentiation is not permanent, in fact many interconversions are possible. Chloroplasts may be converted to chromoplasts, which are pigment-filled plastids responsible for the bright colors seen in flowers and ripe fruit. Starch storing amyloplasts can also be converted to chromoplasts, and it is possible for proplastids to develop straight into chromoplasts. Chromoplasts and amyloplasts can also become chloroplasts, like what happens when a carrot or a potato is illuminated. If a plant is injured, or something else causes a plant cell to revert to a meristematic state, chloroplasts and other plastids can turn back into proplastids. Chloroplast, amyloplast, chromoplast, proplast, etc., are not absolute states—intermediate forms are common.

Division

Most chloroplasts in a photosynthetic cell do not develop directly from proplastids or etioplasts. In fact, a typical shoot meristematic plant cell contains only 7–20 proplastids. These proplastids differentiate into chloroplasts, which divide to create the 30–70 chloroplasts found in a mature photosynthetic plant cell. If the cell divides, chloroplast division provides the additional chloroplasts to partition between the two daughter cells.

In single-celled algae, chloroplast division is the only way new chloroplasts are formed. There is no proplastid differentiation—when an algal cell divides, its chloroplast divides along with it, and each daughter cell receives a mature chloroplast.

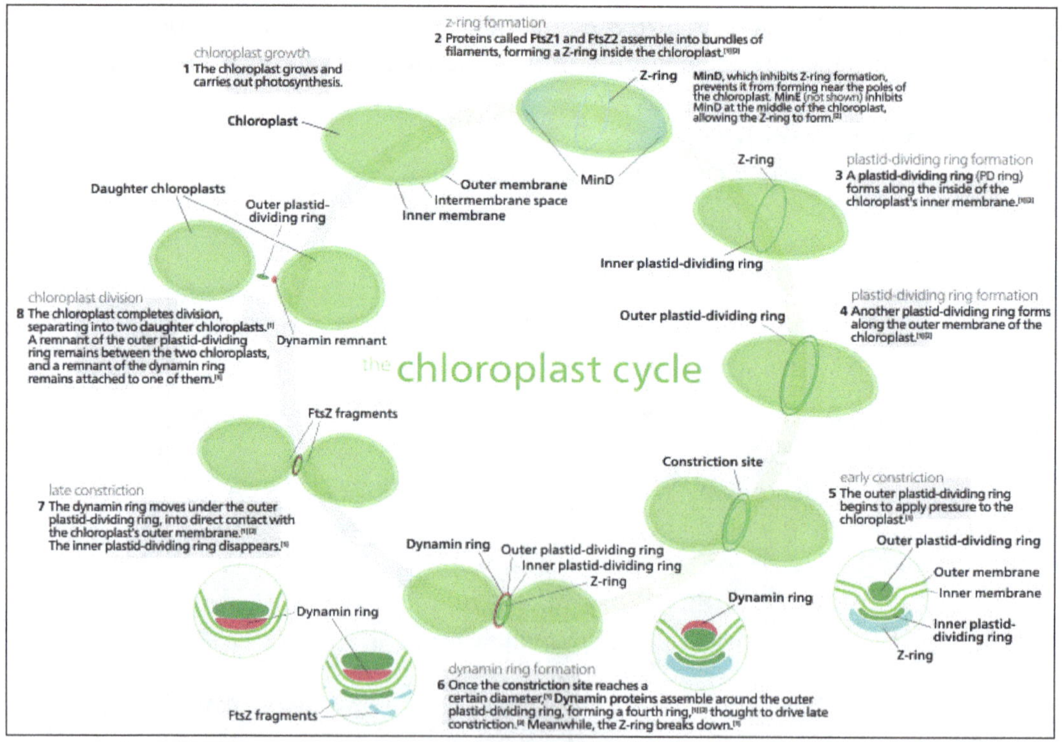

Most chloroplasts in plant cells, and all chloroplasts in algae arise from chloroplast division.

Almost all chloroplasts in a cell divide, rather than a small group of rapidly dividing chloroplasts. Chloroplasts have no definite S-phase—their DNA replication is not synchronized or limited to that of their host cells. Much of what we know about chloroplast division comes from studying organisms like Arabidopsis and the red alga Cyanidioschyzon merolæ.

The division process starts when the proteins FtsZ1 and FtsZ2 assemble into filaments, and with the help of a protein ARC6, form a structure called a Z-ring within the chloroplast's stroma. The Min system manages the placement of the Z-ring, ensuring that the chloroplast is cleaved more or less evenly. The protein MinD prevents FtsZ from linking up and forming filaments. Another protein ARC3 may also be involved, but it is not very well understood. These proteins are active at the poles of the chloroplast, preventing Z-ring formation there, but near the center of the chloroplast, MinE inhibits them, allowing the Z-ring to form.

Next, the two plastid-dividing rings, or PD rings form. The inner plastid-dividing ring is located in the inner side of the chloroplast's inner membrane, and is formed first. The outer plastid-dividing ring is found wrapped around the outer chloroplast membrane. It consists of filaments about 5 nanometers across, arranged in rows 6.4 nanometers apart, and shrinks to squeeze the chloroplast. This is when chloroplast constriction begins. In a few species like Cyanidioschyzon merolæ, chloroplasts have a third plastid-dividing ring located in the chloroplast's intermembrane space.

Late into the constriction phase, dynamin proteins assemble around the outer plastid-dividing ring, helping provide force to squeeze the chloroplast. Meanwhile, the Z-ring and the inner plastid-dividing ring break down. During this stage, the many chloroplast DNA plasmids floating around in the stroma are partitioned and distributed to the two forming daughter chloroplasts.

Later, the dynamins migrate under the outer plastid dividing ring, into direct contact with the chloroplast's outer membrane, to cleave the chloroplast in two daughter chloroplasts.

A remnant of the outer plastid dividing ring remains floating between the two daughter chloroplasts, and a remnant of the dynamin ring remains attached to one of the daughter chloroplasts.

Of the five or six rings involved in chloroplast division, only the outer plastid-dividing ring is present for the entire constriction and division phase—while the Z-ring forms first, constriction does not begin until the outer plastid-dividing ring forms.

Regulation

In species of algae that contain a single chloroplast, regulation of chloroplast division is extremely important to ensure that each daughter cell receives a chloroplast—chloroplasts can't be made from scratch. In organisms like plants, whose cells contain multiple chloroplasts, coordination is looser and less important. It is likely that chloroplast and cell division are somewhat synchronized, though the mechanisms for it are mostly unknown.

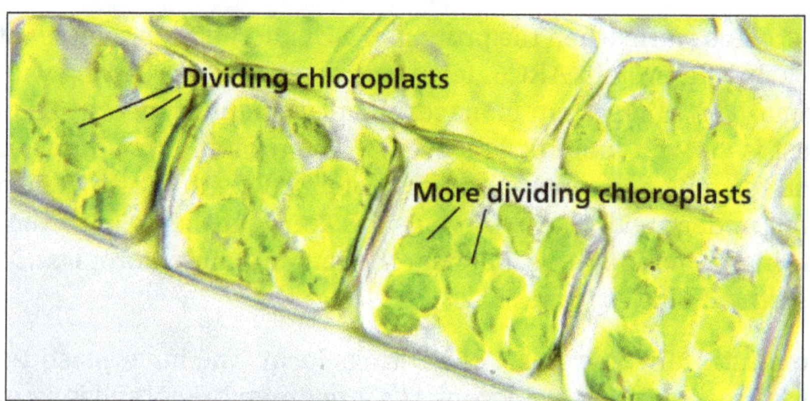

Chloroplast division In this light micrograph of some moss chloroplasts, many dumbbell-shaped chloroplasts can be seen dividing. Grana are also just barely visible as small granules.

Light has been shown to be a requirement for chloroplast division. Chloroplasts can grow and progress through some of the constriction stages under poor quality green light, but are slow to complete division—they require exposure to bright white light to complete division. Spinach leaves grown under green light have been observed to contain many large dumbbell-shaped chloroplasts. Exposure to white light can stimulate these chloroplasts to divide and reduce the population of dumbbell-shaped chloroplasts.

Chloroplast Inheritance

Like mitochondria, chloroplasts are usually inherited from a single parent. Biparental chloroplast inheritance—where plastid genes are inherited from both parent plants—occurs in very low levels in some flowering plants.

Many mechanisms prevent biparental chloroplast DNA inheritance, including selective destruction of chloroplasts or their genes within the gamete or zygote, and chloroplasts from one parent being excluded from the embryo. Parental chloroplasts can be sorted so that only one type is present in each offspring.

Gymnosperms, such as pine trees, mostly pass on chloroplasts paternally, while flowering plants often inherit chloroplasts maternally. Flowering plants were once thought to only inherit chloroplasts maternally. However, there are now many documented cases of angiosperms inheriting chloroplasts paternally.

Angiosperms, which pass on chloroplasts maternally, have many ways to prevent paternal inheritance. Most of them produce sperm cells that do not contain any plastids. There are many other documented mechanisms that prevent paternal inheritance in these flowering plants, such as different rates of chloroplast replication within the embryo.

Among angiosperms, paternal chloroplast inheritance is observed more often in hybrids than in offspring from parents of the same species. This suggests that incompatible hybrid genes might interfere with the mechanisms that prevent paternal inheritance.

Transplastomic Plants

Recently, chloroplasts have caught attention by developers of genetically modified crops. Since, in most flowering plants, chloroplasts are not inherited from the male parent, transgenes in these plastids cannot be disseminated by pollen. This makes plastid transformation a valuable tool for the creation and cultivation of genetically modified plants that are biologically contained, thus posing significantly lower environmental risks. This biological containment strategy is therefore suitable for establishing the co-existence of conventional and organic agriculture. While the reliability of this mechanism has not yet been studied for all relevant crop species, recent results in tobacco plants are promising, showing a failed containment rate of transplastomic plants at 3 in 1,000,000.

Chromoplast

Chromoplasts are plastids, heterogeneous organelles responsible for pigment synthesis and storage in specific photosynthetic eukaryotes. It is thought that like all other plastids including chloroplasts and leucoplasts they are descended from symbiotic prokaryotes.

The coloration of the petals and sepals on the Bee orchid is controlled by a specialized organelle in plant cells called a chromoplast.

Function

Chromoplasts are found in fruits, flowers, roots, and stressed and aging leaves, and are responsible for their distinctive colors. This is always associated with a massive increase in the accumulation of carotenoid pigments. The conversion of chloroplasts to chromoplasts in ripening is a classic example.

They are generally found in mature tissues and are derived from preexisting mature plastids. Fruits and flowers are the most common structures for the biosynthesis of carotenoids, although other reactions occur there as well including the synthesis of sugars, starches, lipids, aromatic compounds, vitamins and hormones. The DNA in chloroplasts and chromoplasts is identical. One subtle difference in DNA was found after a liquid chromatography analysis of tomato chromoplasts was conducted, revealing increased cytosine methylation.

Chromoplasts synthesize and store pigments such as orange carotene, yellow xanthophylls, and various other red pigments. As such, their color varies depending on what pigment they contain. The main evolutionary purpose of chromoplasts is probably to attract pollinators or eaters of colored fruits, which help disperse seeds. However, they are also found in roots such as carrots and sweet potatoes. They allow the accumulation of large quantities of water-insoluble compounds in otherwise watery parts of plants.

When leaves change color in the autumn, it is due to the loss of green chlorophyll, which unmasks preexisting carotenoids. In this case, relatively little new carotenoid is produced—the change in plastid pigments associated with leaf senescence is somewhat different from the active conversion to chromoplasts observed in fruit and flowers.

There are some species of flowering plants that contain little to no carotenoids. In such cases there are plastids present within the petals that closely resemble chromoplasts and are sometimes visually indistinguishable. Anthocyanins and flavonoids located in the cell vacuoles are responsible for other colors of pigment.

The term "chromoplast" is occasionally used to include any plastid that has pigment, mostly to emphasize the difference between them and the various types of leucoplasts, plastids that have no pigments. In this sense, chloroplasts are a specific type of chromoplast. Still, "chromoplast" is more often used to denote plastids with pigments other than chlorophyll.

Structure and Classification

Using a light microscope chromoplasts can be differentiated and are classified into four main types. The first type is composed of proteic stroma with granules. The second is composed of protein crystals and amorphous pigment granules. The third type is composed of protein and pigment crystals. The fourth type is a chromoplast which only contains crystals. An electron microscope reveals even more, allowing for the identification of substructures such as globules, crystals, membranes, fibrils and tubules. The substructures found in chromoplasts are not found in the mature plastid that it divided from.

The presence, frequency and identification of substructures using an electron microscope has led to further classification, dividing chromoplasts into five main categories: Globular chromoplasts, crystalline chromoplasts, fibrillar chromoplasts, tubular chromoplasts and membranous chromoplasts. It has also been found that different types of chromoplasts can coexist in the same organ. Some examples of plants in the various categories include mangoes, which have globular chromoplasts, and carrots which have crystalline chromoplasts.

Although some chromoplasts are easily categorized, others have characteristics from multiple categories that make them hard to place. Tomatoes accumulate carotenoids, mainly lycopene crystalloids in membrane-shaped structures, which could place them in either the crystalline or membranous category.

Evolution

Plastids are descendants of cyanobacteria, photosynthetic prokaryotes, which integrated themselves into the eukaryotic ancestor of algae and plants, forming an endosymbiotic relationship. The ancestors of plastids diversified into a variety of plastid types, including chromoplasts. Plastids also possess their own small genome and some have the ability to produce a percentage of their own proteins.

The main evolutionary purpose of chromoplasts is to attract animals and insects to pollinate their flowers and disperse their seeds. The bright colors often produced by chromoplasts is one of many ways to achieve this. Many plants have evolved symbiotic relationships with a single pollinator. Color can be a very important factor in determining which pollinators visit a flower, as specific colors attract specific pollinators. White flowers tend to attract beetles, bees are most often attracted to violet and blue flowers, and butterflies are often attracted to warmer colors like yellows and oranges.

Leucoplast

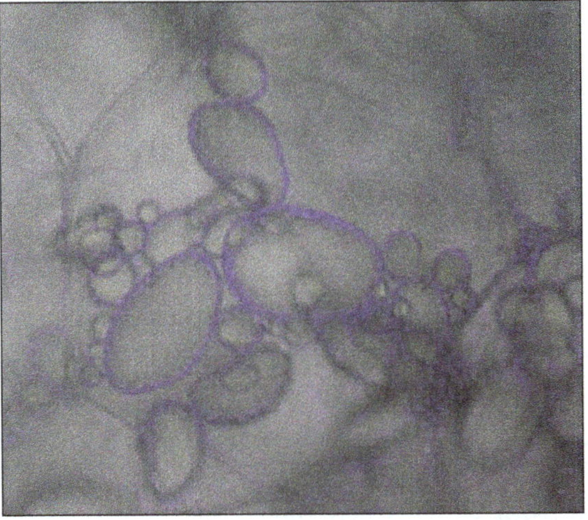

Leucoplasts, specifically, amyloplasts.

Leucoplasts are a category of plastid and as such are organelles found in plant cells. They are non-pigmented, in contrast to other plastids such as the chloroplast.

Lacking photosynthetic pigments, leucoplasts are not green and are located in non-photosynthetic tissues of plants, such as roots, bulbs and seeds. They may be specialized for bulk storage of starch, lipid or protein and are then known as amyloplasts, elaioplasts, or proteinoplasts (also called aleuroplasts) respectively. However, in many cell types, leucoplasts do not have a major storage function and are present to provide a wide range of essential biosynthetic functions, including the synthesis of fatty acids such as palmitic acid, many amino acids, and tetrapyrrole compounds such as heme. In general, leucoplasts are much smaller than chloroplasts and have a variable morphology, often described as amoeboid. Extensive networks of stromules interconnecting leucoplasts have been observed in epidermal cells of roots, hypocotyls, and petals, and in callus and suspension culture cells of tobacco. In some cell types at certain stages of development, leucoplasts are clustered around the nucleus with stromules extending to the cell periphery, as observed for proplastids in the root meristem.

Etioplasts, which are pre-granal, immature chloroplasts but can also be chloroplasts that have been deprived of light, lack active pigment and can be considered leucoplasts. After several minutes exposure to light, etioplasts begin to transform into functioning chloroplasts and cease being leucoplasts. Amyloplasts are of large size and store starch. Proteinoplasts store proteins and are found in seeds (pulses). Elaioplasts store fats and oils and are found in seeds. They are also called oleosomes.

Golgi Apparatus

The plant Golgi apparatus has an important role in protein glycosylation and sorting, but is also a major biosynthetic organelle that synthesises large quantities of cell wall polysaccharides.

Organisation of the Golgi Apparatus

The plant Golgi apparatus is composed of many small stacks of cisternae, sometimes known as dictyosomes. The number of stacks and their distribution within the cell is dependent on the cell type. Maize root cap cells, which are actively secreting large amounts of mucopolysaccharides, contain between 300 and 600 Golgi stacks per cell but there are many fewer, on average 25, in the apical meristem cells of the hairy-willow herb Epilobium hirsutum (and references within). The stacks are usually dispersed throughout the plant cell cytoplasm singly or in small groups. This distribution can be illustrated dramatically by immunofluorescence labelling with the monoclonal antibody JIM84. This antibody recognises a glycoprotein epitope present in the Golgi, and also in the plasma membrane of some species. The dispersed distribution of the stacks may result from continual streaming through the cytoplasm, but a direct visualisation in a living cell to demonstrate this movement has not yet been achieved. A further consequence of cytoplasmic streaming may be a loss of a fixed relationship between the Golgi and the ER.

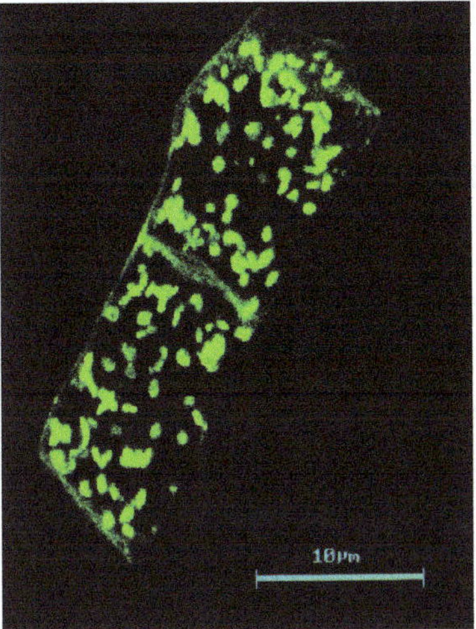

Distribution of Golgi bodies within two Nicotiana root cells shown by labelling with JIM84. The monoclonal antibody JIM84 recognises a carbohydrate epitope on many Golgi and plasma membrane glycoproteins. Golgi bodies are dispersed singly or in small groups throughout the cytoplasm.

In cells with localised sites of cell wall growth or intensive secretion, Golgi stacks can become concentrated to specific subcellular regions. Within tip growing cells, like root hairs and pollen tubes, Golgi bodies are localised in a region a small distance away from the growing tip. Golgi-derived vesicles containing cell wall polysaccharides concentrate with other vesicles in the extreme tip of the growing root hair where most of the other organelles are excluded. However, an association of the Golgi with growing regions of the cell is not absolute. It is also probable that vesicles produced by Golgi stacks a relatively long distance from the wall can be targeted to the site of cell growth.

Movement of secretory vesicles away from the Golgi stacks can be inhibited by cytochalasin B, suggesting that actin microfilaments have a role in vesicle movement and may also have a role in distributing Golgi stacks throughout the cytoplasm. When microtubules are disrupted, vesicles do migrate away from the Golgi stack, but the site of vesicle fusion may be altered, especially in tip-growing cells.

Plant cells divide by the de novo formation of a separating cell wall between daughter cells. Formation of the cell plate requires intensive secretion of Golgi-synthesised cell wall polysaccharides. The phragmoplast, a structure containing microfilaments and microtubules, spatially orients and positions the new cell wall by directing Golgi-derived vesicles to the cell plate. The vesicles align and then fuse in a regulated and characteristic manner. A potential specific t-SNARE that might regulate the fusion process has recently been identified by analysis of a cytokinesis-defective Arabidopsis embryogenesis mutant, knolle. Although Golgi stacks are excluded from the direct area of the forming cell plate, they may associate with the phragmoplast and at later stages become more closely associated with the maturing cell plate. The new plasma membrane, also derived from the Golgi-derived vesicles, acquires typical plasma membrane characteristics as soon as the vesicles fuse, including affinity for the electron microscopy (EM) stain osmium ferricyanide, and the ability to synthesise callose.

The process of cell division highlights a significant difference in behaviour of the Golgi apparatus in plants and animals. To ensure partitioning of membranes during cytokinesis of mammalian cells, secretion stops and the Golgi apparatus vesiculates, reforming after mitosis. In contrast, in plant cells the Golgi apparatus must be intact and functional for cytokinesis to occur. Consequently, the mechanism to partition Golgi stacks between the daughter cells in plants may be unique. It may rely on a dispersed distribution in the cytoplasm before phragmoplast formation, or an interaction with cytoskeletal elements.

The organisation into discrete stacks also has the consequence that their number must increase during the cell cycle, and during differentiation into an elongating or secretory cell. The mechanism of Golgi replication has been addressed by electron microscopic morphological studies. Garcia-Herdugo and co-workers found that the number, but not the size, of Golgi stacks in onion meristematic cells varied during the cell cycle, increasing dramatically during mitosis. Additionally, in a synchronous culture of Madagascar periwinkle (Catharanthus), the size and number of cisternae increased shortly before and during cytokinesis. Golgi stacks divide during cytokinesis by the fission of the individual cisternae, moving in the cis to trans direction. However, it is difficult at present to exclude other mechanisms, including de novo formation of stacks. Perhaps with the discovery of Golgi proteins that can be fused to green fluorescent protein, the timing and mechanism of Golgi replication may be investigated more thoroughly within living plant tissues.

Morphology and Polarisation of Individual Golgi Stacks

The size and morphology of individual plant Golgi stacks vary tremendously between different cell types and species. The cisternal membranes and associated vesicles can be from 0.5 to 2.0 μm in diameter. Golgi stacks are usually composed of three to eight cisternae, and may be composed of 15 or more in scale-secreting algal cells. Nevertheless, within a single cell, the cisternal number is remarkably uniform. Three-dimensional reconstructions of serial EM sections through plant Golgi stacks have provided a morphological view rather different from the common 'stacked-pancake' model. A model of a single dictyosome from Aptenia cordifolia showed that the stack and associated vesicles have an overall spherical shape. The models also revealed that there can be tubular connections between adjacent cisternae, between neighbouring Golgi stacks, and possibly connections to the endoplasmic reticulum.

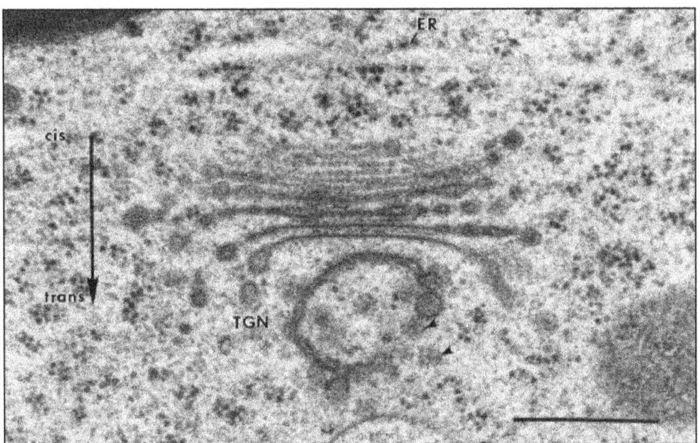

Golgi stack and trans Golgi network (TGN) in a columella cell of a Nicotiana root cap. Morphological characteristics of the cisternae change in a cis to trans direction. Clathrin coats (arrowheads) are associated with the TGN. ER, endoplasmic reticulum. Bar equals 0.5 μm.

The directional flow of material through the Golgi apparatus is reflected in the morphological polarisation of individual Golgi stacks. There are more vesicular profiles associated with the medial region and trans side of the Golgi than the cis cisternae. The cisternae vary in staining characteristics and luminal thickness. The membranes and contents of the different cisternae stain more intensely in a cis to trans direction. Furthermore, the thickness of the cisternal lumen decreases in a cis to trans direction. Because this is more constant in animal Golgi stacks, it was originally believed that the differences in plant cisternae were artefacts of chemical fixation. However, studies of tissues preserved by high-pressure freeze fixation and freeze substitution have confirmed these differences. So-called intercisternal elements can be seen between cisternae on the trans side of the stack in cells of some species, and may be involved in maintaining stack integrity.

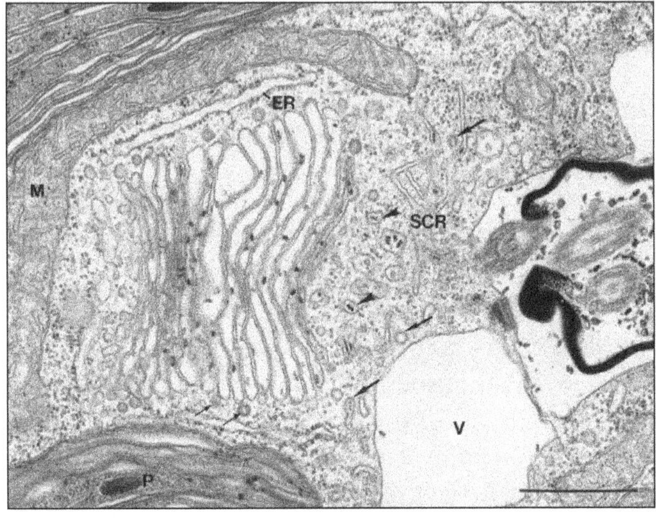

Golgi stack in the alga Scherffelia dubia. A membranous scale reticulum (SCR), and secretory vesicles containing scales (arrowheads) are seen near the trans side of the

Golgi body. Vesicles not containing scales (small arrows) are closely associated with the Golgi stack. Large arrows indicate clathrin-coated membrane profiles. ER, endoplasmic reticulum; M, mitochondrion; P, plastid; V, contractile vacuole. Bar equals 0.5 µm.

In cells secreting large amounts of material, the stack polarity is exaggerated. For example, in slime-secreting maize root cap cells the trans-most cisternal rims are dilated or connected to huge vesicles which bud from the trans side of the Golgi stack. Similar differentiation of Golgi stack morphology is also seen in slime secreting cells of root tip cells of tobacco and Arabidopsis plants. These cells show beautifully how the Golgi morphology of different plant cells reflects their functional differences.

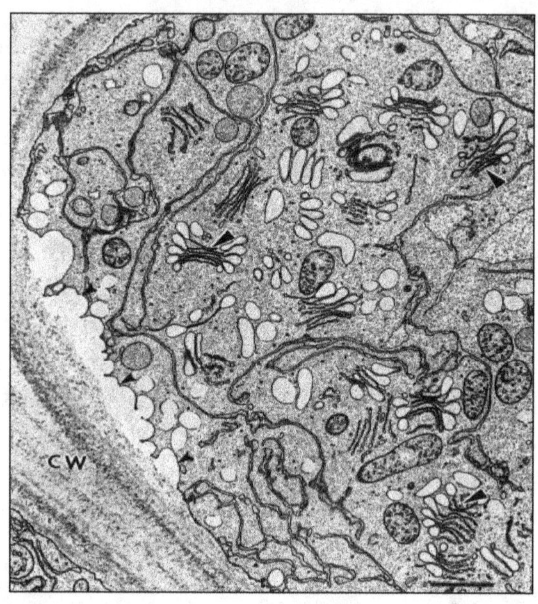

Hyper-secretory outer cap cell from a maize root fixed in potassium permanganate to stain endomembranes. Golgi bodies (large arrowheads) have extended secretory vesicles attached to the cisternae. Small arrowheads depict secreted mucopolysaccharides. CW, cell wall. Bar equals 1 µm.

In many higher plants, there is no obvious cis Golgi network (CGN). In some algae the ER and Golgi are clearly associated and the cis Golgi can be identified. However, in most higher plants, especially in cells secreting polysaccharide, there is no fixed association with the ER. Although this might be considered important for transit of proteins between the two organelles, an association is not always obvious from electron micrographs. Golgi bodies may associate more closely with ER under conditions of cold stress when polysaccharide secretion is reduced. Even in cases where the ER is nearby, membranes that could constitute a CGN are not often apparent.

On the trans side of the stack, slightly removed from it, is a membrane structure referred to as the trans Golgi network (TGN). It is less extensive than in animal cells, and is often reminiscent of a Golgi cisterna. It can be distinguished by its more complex morphology, and by the presence of clathrin-coated and other budding vesicles. EM sectioning has shown that not every stack has a clear TGN and in this case the trans-most cisterna may function in its place. Alternatively, the so-called partially coated reticulum (PCR), a membrane of similar morphology, could be a 'displaced' TGN. Although it is not so closely associated with the stack, it can be connected to it by tubules, and it also contains clathrin-like coat profiles. However, the PCR also has some characteristics of an endosome. Therefore it is not clear if the TGN/PCR is one and the same organelle with functionally distinct subdomains. Alternatively there may be two clearly distinct but morphologically similar organelles. At present, there are no TGN or PCR marker proteins that can be used to distinguish these possibilities.

Golgi morphology, vesicle trafficking and secretion have been studied using Brefeldin A (BFA), a fungal toxin from the plant pathogen Alternaria carthami. The effects of BFA depend on the cell type and experimental conditions. BFA inhibits secretion of glycoproteins and polysaccharides, and disrupts the delivery of soluble proteins to the vacuole. Inhibition of trafficking is not always accompanied by disintegration of the Golgi stack as seen in mammalian cells. In certain cases, Golgi stack morphology and distribution are altered dramatically by BFA and 7-oxo-BFA causing Golgi clustering and vesiculation without fusion with the ER. This further illustrates the more distant relationship between the ER and Golgi in plants than seen in mammalian cells.

Biosynthetic Functions of the Plant Golgi Apparatus

The plant Golgi apparatus synthesises a wide range of cell wall polysaccharides and proteoglycans, and also carries out O-linked glycosylation and N-linked glycan processing. No studies have yet addressed whether glycolipids are modified in the plant Golgi. As in the animal and yeast Golgi, it is thought that a specific glycosyltransferase is required to transfer each sugar from a nucleotide sugar donor to an oligosaccharide acceptor, and the transferases may work in complexes. Although the activities of several of the enzymes have been extensively characterised, there are no published reports of the cloning of any genes encoding them. Transporters are also required to permit entry of the sugar nucleotide precursors into the lumen of the Golgi. The diverse biosynthetic processes therefore require a very large number of specific enzymes. In order to simplify the recognition of the substrates, it is probable that the enzymes are subcompartmentalised within the stack.

The synthesis of cell wall polysaccharides in the Golgi apparatus was demonstrated many years ago by in vivo labelling studies followed by autoradiography. Only cellulose (β1,4-glucan) and callose (β1,3-glucan), are synthesised at the plasma membrane. The two abundant classes of Golgi-synthesised polysaccharides, the pectins and hemicelluloses, can constitute between 50 and 80% of the dry weight of the cell wall. Their structures are complex, containing a range of sugars in a variety of linkages.

The pectins and hemicelluloses are distinguished by their sugar composition and consequent extractibility from the cell wall, the pectin being soluble in hot dilute acid, the hemicelluloses in alkali. The pectins are highly branched acidic polysaccharides composed mainly of galactose and rhamnose, but can contain over ten other sugars. A major pectin of dicotyledonous plants, rhamnogalacturonan I (RG-I), has a backbone of α1,4-linked galacturonic acid alternating with α1,2-linked rhamnose, and a variety of side chains composed principally of arabinose and galactose. The galacturonic acid of pectin is often partially esterified in the Golgi. Other pectins, including arabinans, galactans, arabinogalactans and polygalacturonic acid (PGA) are present in variable amounts in different tissues and species.

The hemicelluloses are a broad group of polymers including xylans, glucans, mannans, glucomannans and galactomannans. The principle hemicellulose of the primary walls of dicotyledonous plants, xyloglucan, has a β1,4-glucan backbone with xylose side chains further modified by galactose and sometimes fucose. In contrast, the secondary walls contain a xylan with side chains of glucuronic acid.

A single Golgi stack can synthesise both pectins and hemicelluloses. The plant Golgi apparatus therefore contains a very wide range of enzymes. The synthesis is highly developmentally regulated, and so the activities are remodelled during cell differentiation. Nevertheless, even within a single cell, several different polysaccharides are synthesised at one time. Understanding the specificity of the enzymes, and the mechanism of their regulation is one of the major challenges of plant Golgi research.

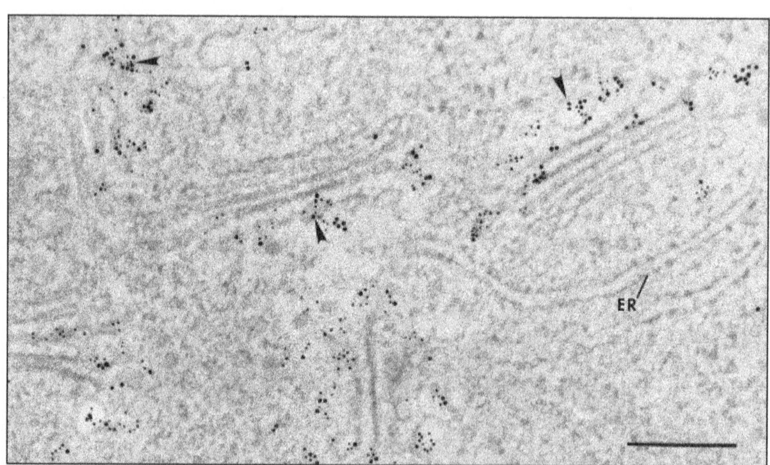

Golgi stack in a root hair of *Vicia faba*. The tissue is immunolabelled to show xyloglucan (large gold) and methyl esterified pectin (small gold) asymmetrically localised within the Golgi stack. Bar equals 0.25 μm.

Glycosylation of the cell wall proteoglycans constitutes a significant proportion of plant Golgi sugar transferase activity. The cell wall arabinogalactan (AGP) and extensin superfamilies of repetitive hydroxyproline-rich glycoproteins (HRGPs) are extensively glycosylated, to the extent that carbohydrate can comprise from 5% to over 90% of their mass. Cell fractionation studies suggest that incorporation of arabinose into

HRGPs occurs predominantly in the Golgi apparatus. Consistent with this biochemical evidence, immunolabelling with antibody raised against hydroxyproline-arabinosylated extensin labels the Golgi but not the ER. Hydroxyproline (Hyp) of HRGPs can be the site of attachment of three to four arabinose residues, and serine may bear a single galactose. Since not all Hyp residues are equally glycosylated, recent studies have investigated the recognition motifs of the glycosyltransferases. In a Douglas fir HRGP, the peptide Lys-Pro-Hyp-Hyp-Val is always arabinosylated at Hyp-3, mainly by a triarabinoside. In general, blocks of hydroxyproline are the most highly arabinosylated, single non-contiguous Hyp residues the least, but single Hyp residues are likely attachment sites for the larger arabinogalactan glycan structures of AGPs.

In general, O-linked glycosylation of intracellular proteins seems to be more limited, but has been little studied. Sporamin, a sweet potato vacuolar storage protein becomes glycosylated on hydroxyproline, and perhaps serine. Sensitivity to BFA and pulse chase experiments suggest this glycosylation occurs in the Golgi. Processing of N-linked glycans is similar to that in animals. A significant difference is the presence of $\alpha1,3$-linked fucose (and not $\alpha1,6$-linked fucose) and a $\beta1,2$-xylose linked to the Man3GlcNAc2 core. As a consequence, many plant glycoproteins are highly antigenic. There is some heterogeneity in further processing, with glycans of some proteins receiving galactose and fucose to form the Lewis a antigen. Sialic acid has not been detected in plant proteins. The activities of the glycosidases and glycosyltransferases have been localised to the Golgi apparatus by cell fractionation.

Biochemical Subcompartmentation of the Golgi

The complexity of the glycosylation processes in the plant Golgi, particularly polysaccharide synthesis, has led to the suggestion that biochemical subcompartmentation of the enzymes into cis, medial and trans cisternae would simplify the substrate recognition and synthetic mechanisms. Although there are probably three distinct compartments in the mammalian and yeast Golgi complex, and it is possible to define morphologically three types of cisternae, the functional equivalence of these compartments in plants is unproven. There has been no unequivocal demonstration of biochemical compartmentation of enzymes within the plant Golgi stack. The enzymes have not been localised microscopically, since there are no antibodies or clones yet available for these purposes. Furthermore, the absence of a clear CGN and TGN in certain cells, and the morphology suggestive of a maturing stack of cisternae, have led to the suggestion that the plant stack might be a single compartment, analogous to maturing secretory vesicles. It is important to resolve this point, since it has implications for targeting of the Golgi enzymes, for understanding the mechanism of movement of proteins and polysaccharides through the stack, and for the mechanism and specificity of polysaccharide biosynthesis.

There is considerable indirect evidence that the enzyme activities are compartmentalised into distinct cisternae, and morphological polarity would support the idea. In a series of immunocytochemical experiments, Staehelin and co-workers, and others, have shown that the polysaccharide products of the Golgi enzymes are asymmetrically distributed across the stack. Using antisera that recognised the xyloglucan backbone, labelling was found over the trans cisternae and TGN. A similar distribution of labelling was found with antibody recognising a fucose-containing epitope of the xyloglucan side chain. Moreover, antibodies recognising epitopes of the pectin PGA/RG-I labelled predominantly the cis and medial cisternae in cultured sycamore cells. Methyl esterified pectin is found in the medial and trans cisternae of vetch root hairs and sycamore cells. This suggests that pectin synthesis is initiated in the cis Golgi and methyl esterified in later cisternae, whereas the hemicellulose is synthesised predominantly in the trans cisternae. However, it is unclear whether all precursor and final processed polysaccharides are quantitatively recognised by these antibodies. These studies can only show the steady state distribution of product, and not of the enzymes themselves. Recently, a reversibly glycosylated peripheral protein RGP1, suggested to be involved in polysaccharide synthesis, has been shown to be localised specifically to the cytosolic side of cisternae on the trans side of the Golgi stack, further suggesting a specific biochemical function for these cisternae.

Similar immunocytochemical studies using antisera recognising complex N-linked glycans are also consistent with the model of subcompartmentation of enzymatic activities. In addition to labelling of post-Golgi compartments, labelling for xylose-containing glycoproteins was highest in the medial cisternae, while fucose-containing glycoproteins were mainly detected in the trans cisternae. Immunolocalisation of the complex N-linked glycan Lewis a epitope also revealed a late-Golgi labelling. This could suggest that xylosylation of glycoproteins occurs earlier in the stack than fucosylation. However, labelling could be found even in the cis cisternae, and in an earlier study, little evidence for an symmetric distribution was found. Interpretation of these studies is affected by potential recycling of glycoproteins within the stack.

Many studies have attempted to demonstrate subcompartmentation of enzymes by biochemical fractionation. However, the only universally used marker activity is latent UDPase, a Golgi enzyme activity thought to be required for nucleotide-sugar incorporation. There are no general marker activities for early or late Golgi compartments. Homogenates of tissues separated by density gradient centrifugation have separated latent UDPase into two or three distinct membrane fractions. This may reflect separation of subcompartments of different densities. Furthermore, in experiments looking at hemicellulose synthesis, a glucuronyl transferase and a xylosyltransferase activity from pea epicotyl were separated into membranes of different densities, indirectly suggesting distinct subcompartmentation of the enzymes. However, the results could also represent separation of Golgi stacks from different cell types synthesising the different polysaccharides of the primary and secondary cell walls. In contrast to the results suggesting separation of subcompartments, the separation of the enzymes that process N-linked glycans could not be achieved.

Sorting Functions of the Plant Golgi Apparatus

The Golgi is involved in returning escaped proteins back to the ER, sorting of proteins and polysaccharides to the cell wall or vacuoles, and in organising the compartmentation of its own enzymes by retention or retrieval mechanisms. The evidence for compartmentation of glycosyltransferases suggests that they are effectively maintained in the plant Golgi stack. There are few reports yet of studies addressing the question of targeting to the Golgi, because of the absence of identified plant Golgi membrane proteins. A GlcNAc transferase I mutant cgl1 of Arabidopsis could be complemented by expression of the mammalian homologue, suggesting that targeting may be conserved to some extent between mammals and plants. Transgene transferase activity was confined to Golgi-enriched biochemical fractions, but it cannot be ruled out that inactive protein was present elsewhere in the cells.

Since the initial electron microscopic characterisation of the plant Golgi, there has been considerable discussion about the mechanism of movement of polysaccharides and cell wall material through the stack. The two favoured models are cisternal progression and COPI (coatamer)-mediated vesicular transport, and the relative merits of the models have recently been well reviewed. The earlier cisternal progression model is supported by observations of the wide range of unicellular algae that synthesise scales. These scales form within the cisternae, and are never present in Golgi associated vesicles. Maturation of cisternae into secretory vesicles containing large scales can sometimes be directly observed. Although these dramatic examples come from algae, the progression model is also attractive in higher plants. The morphology of the cisternae and the TGN are sometimes very similar. Moreover, in slime secreting cells, the material transported across the stack is not seen in associated coated vesicles. However, improvements to the original progression model are needed to explain the maintenance of compartmentation of enzymes across the stack, and the regeneration of the cis cisternae, given the spatial separation of the ER from the stack.

Escaped ER proteins must be retrieved and returned to the ER. The signal for retrieval of ER luminal proteins is a C-terminal KDEL or HDEL ; retrieval of membrane proteins has not yet been studied in plants. There is no clear structure like a CGN near many Golgi stacks, and there is often no association of the ER and Golgi stack. The location of the retrieval compartment is therefore a matter of some debate. The compartment for retrieval of the escaped ER proteins may not be part of the stack itself, but a discrete compartment with an ER association. Nevertheless, possible COPI coat profiles are sometimes seen on Golgi cisternae, suggesting that some COPI-mediated retrieval occurs within the stack. A probable receptor for retrieval of luminal ER proteins, aERD2p, has been identified, and it will be important to determine its location in the plant cell to demonstrate the site of the retrieval compartment.

There are several requirements for sorting processes at the trans face of the Golgi apparatus. The vacuolar proteins are sorted from the proteins destined for the plasma membrane or cell wall. Polysaccharides are also excluded from vesicles that take proteins to the vacuoles. Furthermore, the recent demonstration that some plant cells have two distinct vacuoles, a lytic and a storage organelle, raises the possibility that the proteins destined for these organelles are sorted from each other in the Golgi. There are two candidate vesicle types seen at the TGN that might be involved. Clathrin coated vesicles are probably involved in transport of vacuolar proteins in some cells. However, in the developing seed storage tissues of pea cotyledons and pumpkin, smooth dense vesicles (SDVs) carrying seed storage proteins bud from the TGN. These SDVs, up to 300 nm in diameter, have no apparent coat. It is not yet clear if these different vesicle types are generally involved in transport to the different vacuoles. Several studies have identified peptide signals that are able to target test proteins to the vacuole. These peptide signals, present as propeptides, may be at the N or C terminus of the protein, and are poorly conserved. A probable vacuolar sorting receptor that recognises the N-terminal propeptide of aleurain, a barley storage protein, has recently been identified and localised by immunoelectron microscopy to the Golgi stack and a putative prevacuolar compartment. Wortmannin can inhibit the sorting of a C-terminally targeted protein, barley lectin, suggesting the involvement of a PI kinase in the sorting process. However, the same protein targeted with an N-terminal sequence from sweet potato sporamin was relatively insensitive. This supports the idea that there are two different sorting pathways, and it is tempting to speculate that the different pathways and peptide signals are related to the need to target to different vacuole types.

Endoplasmic Reticulum

The endoplasmic reticulum (ER) is a large organelle made of membranous sheets and tubules that begin near the nucleus and extend across the cell. The endoplasmic reticulum creates, packages, and secretes many of the products created by a cell. Ribosomes, which create proteins, line a portion of the endoplasmic reticulum.

The entire structure can account for a large proportion of the endomembrane system of the cell. For instance, in cells such as liver hepatocytes that are specialized for protein secretion and detoxification, the ER can account for more than 50% of the total lipid bilayer of the cell. Similarly, the ER membrane system is particularly prominent in pancreatic beta cells that secrete insulin, or within activated B-lymphocytes that produce antibodies.

As seen in the image, the membranes of the endoplasmic reticulum are contiguous with the outer nuclear membrane, even though their compositions can be different. The ER contains special membrane-embedded proteins that stabilize its structure and curvature. This organelle acts as an important regulator of cell function because it interacts closely with a number of other organelles. Products of the endoplasmic reticulum often travel to the Golgi body for packaging and additional processing before being secreted.

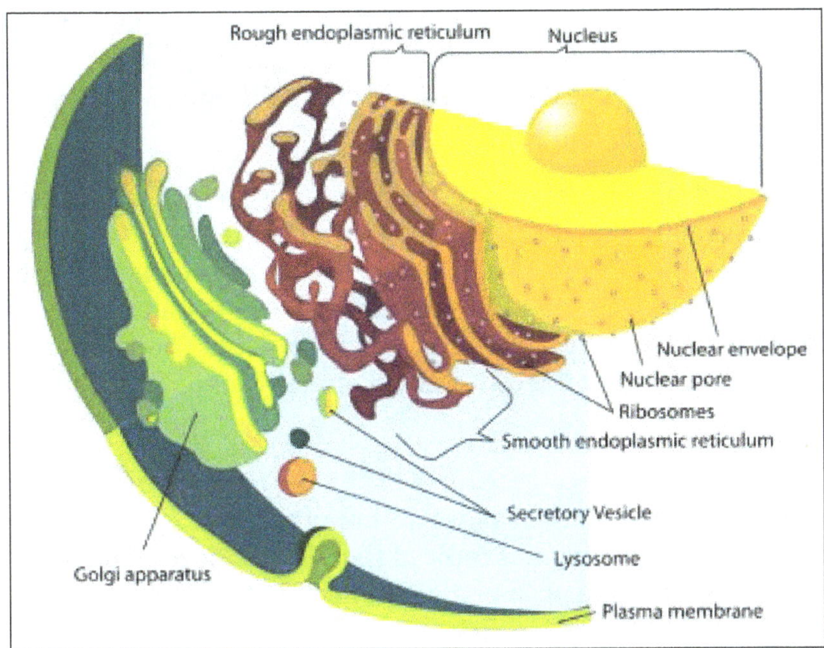

Endomembrane system diagram.

Endoplasmic Reticulum Function

The ER plays a number of roles within the cell, from protein synthesis and lipid metabolism to detoxification of the cell. Cisternae, each of the small folds of the endoplasmic reticulum, are commonly associated with lipid metabolism. This creates the plasma membrane of the cell, as well as additional endoplasmic reticulum and organelles. They also appear to be important in maintaining the Ca^{2+} balance within the cell and in the interaction of the ER with mitochondria. This interaction also influences the aerobic status of the cell.

ER sheets appear to be crucial in the response of the organelle to stress, especially since cells alter their tubules-to-sheets ratio when the number of unfolded proteins increases. Occasionally, apoptosis is induced by the ER in response to an excess of unfolded protein within the cell. When ribosomes detach from ER sheets, these structures can disperse and form tubular cisternae.

Although ER sheets and tubules appear to have distinct functions, there isn't a perfect delineation of roles. For instance, in mammals tubules and sheets can interconvert, making the cells adaptable to various conditions. The relationship between structure and function in the ER has not been completely elucidated.

Protein Synthesis and Folding

Protein synthesis occurs in the rough endoplasmic reticulum. Although translation for all proteins begins in the cytoplasm, some are moved into the ER in order to be folded and sorted for different destinations. Proteins that are translocated into the ER during translation are often destined for secretion. Initially, these proteins are folded within the ER and then moved into the Golgi apparatus where they can be dispatched towards other organelles.

For instance, the hydrolytic enzymes in the lysosome are generated in this manner. Alternately, these proteins could be secreted from the cell. This is the origin of the enzymes of the digestive tract. The third potential role for proteins translated in the ER is to remain within the endomembrane system itself. This is particularly true for chaperone proteins that assist in the folding of other proteins. The genes encoding these proteins are upregulated when the cell is under stress from unfolded proteins.

Lipid Synthesis

The smooth endoplasmic reticulum plays an important role in cholesterol and phospholipid biosynthesis. Therefore, this topic of the ER is important not only for the generation and maintenance of the plasma membrane but of the extensive endomembrane system of the ER itself.

The SER is enriched in enzymes involved in sterol and steroid biosynthetic pathways and is also necessary for the synthesis of steroid hormones. Therefore the SER is extremely prominent in the cells of the adrenal gland that secrete five different groups of steroid hormones that influence the metabolism of the entire body. The synthesis of these hormones also involves enzymes within the mitochondria, further underscoring the relationship between these two organelles.

Calcium Store

The SER is an important site for the storage and release of calcium in the cell. A modified form of the SER called sarcoplasmic reticulum forms an extensive network in contractile cells such as muscle fibers. Calcium ions are also involved in the regulation of metabolism in the cell and can change cytoskeletal dynamics.

The extensive nature of the ER network allows it to interact with the plasma membrane and use Ca^{2+} for signal transduction and modulation of nuclear activity. In association with mitochondria, the ER can also use its calcium stores to induce apoptosis in response to stress.

Structure of the Endoplasmic Reticulum

The endoplasmic reticulum membrane system can be morphologically divided into two structures—cisternae and sheets. Cisternae are tubular in structure and form a three-dimensional polygonal network. They are about 50 nm in diameter in mammals and 30 nm in diameter in yeast. ER sheets, on the other hand, are membrane-enclosed, two-dimensional flattened sacs that extend across the cytoplasm. They are frequently associated with ribosomes and special proteins called translocons that are necessary for protein translation within the RER.

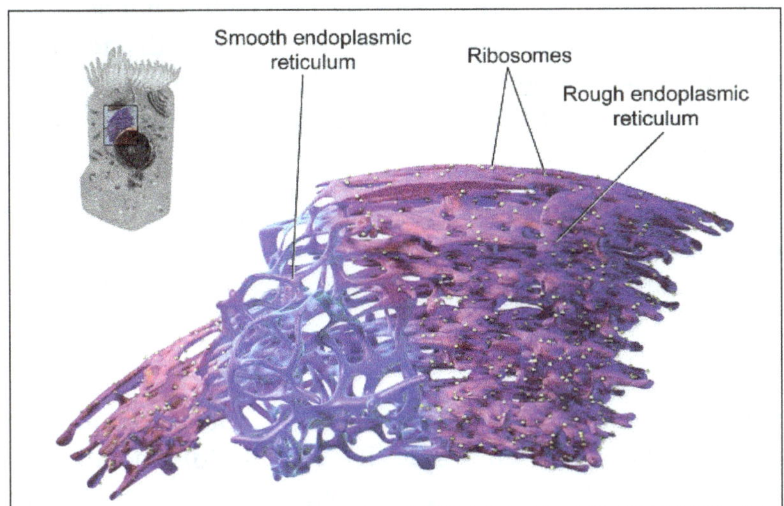

Endoplasmic Reticulum Structure.

The high-curvature of ER tubules is stabilized by the presence of proteins called reticulons and DP1/Yop1p. Reticulons are membrane-associated proteins encoded by four genes in mammals (RTN1-4). These proteins localize to ER tubules and the curved edges of ER sheets. DP1/Yop1p are a class of integral membrane proteins involved in stabilizing the structure of ER cisternae.

Both reticulons and DP1/Yop1 proteins form oligomers and interact with the cytoskeleton. Oligomerization seems to be one of the mechanisms used by these proteins to shape the lipid bilayer into a tubule. Additionally, they also appear to use a wedge-like structural motif that causes the membrane to curve. These two classes of proteins are redundant, since the overexpression of one protein can compensate for the lack of the other protein.

The construction of the ER is intimately involved with the presence of cytoskeletal elements, especially microtubules. ER membranes, especially cisternae, move and branch along microtubules. When microtubule structure is temporarily disrupted, the ER network collapses and reforms only after the microtubule cytoskeleton is reestablished. In addition, changes to the pattern of microtubule polymerization are reflected in changes to ER morphology.

Endoplasmic Reticulum Location

The endoplasmic reticulum processes most of the instructions from the nucleus. As such, the endoplasmic reticulum surrounds the nucleus and radiates outward. In cells that secrete many products for the rest of the body, the endoplasmic reticulum can account for more than 50% of the cell.

In general, the nucleus expresses mRNA (messenger RNA), which tells the cell how to build proteins. The rough endoplasmic reticulum has many ribosomes, which are the primary location of protein production. This portion of the organelle creates proteins and begins to fold them into the proper formation. The smooth endoplasmic reticulum is the primary location for lipid synthesis. As such, it does not contain any ribosomes. Rather, it conducts a series of reactions which create the phospholipid molecules necessary to create various membranes and organelles.

The rough version of the endoplasmic reticulum is often closer to the nucleus, whereas the smooth endoplasmic reticulum is further from the nucleus. However, both versions are connected to each other and the nucleus through a series of small tubules.

Types of Endoplasmic Reticulum

There are two major types of ER within each cell – smooth endoplasmic reticulum (SER) and rough endoplasmic reticulum (RER). Each has distinct functions, and often, differing morphology. The SER is involved in lipid metabolism and acts as the calcium store for the cell. This is particularly important in muscle cells that need Ca^{2+} ions for contraction. The SER is also involved in the synthesis of phospholipids and cholesterol. It is often located near the periphery of the cell.

On the other hand, the RER is commonly seen close to the nucleus. It contains membrane-bound ribosomes that give it the characteristic 'rough' appearance. These ribosomes are creating proteins that are destined for the lumen of the ER and are moved into the organelle as they are being translated. These proteins contain a short signal created by a few amino acids in their N-terminus and are initially translated in the cytoplasm.

However, as soon as the signal is translated, special proteins bind to the growing polypeptide chain and move the entire ribosome and associated translation machinery to the ER. These polypeptides could be resident proteins of the RER, or be moved towards the Golgi network to be sorted and secreted.

Phragmoplast

The phragmoplast is a plant cell specific structure that forms during late cytokinesis. It serves as a scaffold for cell plate assembly and subsequent formation of a new cell wall separating the two daughter cells. The phragmoplast can only be observed in Phragmoplastophyta, a clade that includes the Coleochaetophyceae, Zygnematophyceae, Mesotaeniaceae, and Embryophyta (land plants). Some algae use another type of microtubule array, a phycoplast, during cytokinesis.

Structure

The phragmoplast is a complex assembly of microtubules (MTs), microfilaments (MFs), and endoplasmic reticulum (ER) elements, that assemble in two opposing sets perpendicular to the plane of the future cell plate during anaphase and telophase. It is initially barrel-shaped and forms from the mitotic spindle between the two daughter nuclei while nuclear envelopes reassemble around them. The cell plate initially forms as a disc between the two halves of the phragmoplast structure. While new cell plate material is added to the edges of the growing plate, the phragmoplast microtubules disappear in the center and regenerate at the edges of the growing cell plate. The two structures grow outwards until they reach the outer wall of the dividing cell. If a phragmosome was present in the cell, the phragmoplast and cell plate will grow through the space occupied by the phragmosome. They will reach the parent cell wall exactly at the position formerly occupied by the preprophase band.

The microtubules and actin filaments within the phragmoplast serve to guide vesicles with cell wall material to the growing cell plate. Actin filaments are also possibly involved in guiding the phragmoplast to the site of the former preprophase band location at the parent cell wall. While the cell plate is growing, segments of smooth endoplasmic reticulum are trapped within it, later forming the plasmodesmata connecting the two daughter cells.

The phragmoplast can be differentiated topographically into two areas, the midline that includes the central plane where some of the plus-ends of both anti-parallel sets of microtubules (MTs) interdigitate (as in the midbody matrix), and the distal regions at both sides of the midline.

Role in the Plant Cell Cycle

After anaphase, the phragmoplast emerges from the remnant spindle MTs in between the daughter nuclei. MT plus ends overlap the equator of phragmoplast at the site where the cell plate will form. The formation of the cell plate depends on localized secretory vesicle fusion to deliver membrane and cell-wall components. Excess membrane lipid and cell-wall components are recycled by clathrin/dynamin-dependent retrograde membrane traffic. Once the initial cell plate forms at its center, the phragmoplast begins to expand outward to reach the cell edges. Actin filaments also localize to phragmoplast and accumulate greatly at late telophase. Evidence suggests that actin filaments serve phragmoplast expansion more than initial organization, given that disorganization of actin filaments via drug treatments lead to the delay of cell-plate expansion.

Many microtubule-associated proteins (MAPs) have been localized to the phragmoplast, including both constitutively expressed ones (such as MOR1, katanin, CLASP, SPR2, and γ-tubulin complex proteins) and those expressed specifically during M-phase, such as EB1c, TANGLED1 and augmin complex proteins. The functions of these proteins in the phragmoplast are presumably similar to their functions elsewhere in the cell. Most research into phragmoplast MAPs have been focused on the midline because it is, first, where most of the membrane fusion takes place and, second, where the two sets of anti-parallel MTs are held together. The discovery of an important variety of molecules that localize to the phragmoplast midline is shedding light on the complex processes operating in this phragmoplast region.

Two proteins that have critical functions for antiparallel MT bundling at the phragmoplast midline are MAP65-3 and kinesin-5. The kinesin-7 family proteins, HINKEL/AtNACK1 and AtNACK2/TES, recruit a mitogen-activated protein kinase (MAPK) cascade to the midline and induce MAP65 phosphorylation. Phosphorylated MAP65-1 also accumulates at the midline and reduces MT-bundling activities for cell-plate expansion. The essential mechanism of MAPK cascade for phragmoplast expansion is suppressed by cyclin dependent kinase (CDK) activity before telophase.

Certain phragmoplast midline-accumulating MAPs are essential proteins for cytokinesis. The kinesin-12 members, PAKRP1 and PAKRP1L, accumulate at the midline and double loss-of-function mutants have defective cytokinesis during male gametogenesis. PAKRP2 accumulates at midline and also in puncta throughout the phragmoplast, which implies that PAKRP2 participates in Golgi-derived vesicle transport. Moss homologs of PAKRP2, KINID1a, and KINID1b localize to the phragmoplast midline and are essential for phragmoplast organization. RUNKEL, which is a HEAT repeat-containing MAP, also accumulates at the midline and cytokinesis is aberrant in lines with the loss-of-function mutations in this protein. Another midline-localized protein, "two-in-on" (TIO), is a putative kinase and is also required for cytokinesis as shown by defects in a mutant. TIO interacts with PAKRP1, PAKRP1L (kinesin-12), and NACK2/TES (kinesin-7) according to the yeast two hybrid assays. Finally, TPLATE, an adaptin-like protein, accumulates at the cell plate and is essential for cytokinesis.

Plant Tissue

Plant tissue is a collection of similar cells performing an organized function for the plant. Each plant tissue is specialized for a unique purpose, and can be combined with other tissues to create organs such as leaves, flowers, stems and roots. The following is a brief outline of plant tissues, and their functions within the plant.

Types of Tissue in Plants

Simple Plant Tissue

There are several basic forms of plant tissue, formed from mostly identical types of cells. The first is the epidermis. The epidermis in plants serves the same function as it does in animals. It is a plant tissue formed of thin and densely packed cells, meant to separate the inside of the organisms from the outside. The epidermis is often covered in a layer of waxy protection, to stop the plant from burning or drying out in the sun. The epidermis also contains guard cells, which operate small opening called stoma. These stoma control the passage of air and water through the leaves, allowing plants to move water and nutrients up from the soil.

Sometimes, another form of simple plant tissues covers the epidermis, cork. Cork is a plant tissue seen in woody plants, which dies and becomes an outer layer of bark. This tissue is also soaked with a special waxy substance which protects against insects, the sun, and the elements.

As you turn inside the plants, the next plant tissue is parenchyma. This tissue is comprised of thin-walled cells with very large central vacuoles. The turgor pressure of these vacuoles is elevated when they are full of water, which gives structure and support to the plant. Parenchyma plant tissue is found in all parts of the plant, and makes up large portions of the leaves, stems and roots. In the leaves, parenchyma plant tissue is highly involved in the process of photosynthesis. All parenchyma plant tissue is living, and carries out functions continually. Parenchyma tissue, when wounded, can revert back into meristematic plant tissue to regrow damaged areas.

Like cork, sclerenchyma plant tissue is a structural tissue which dies, but the cell wall and structure remain. Sclerenchyma plant tissue forms long, connected fibers called sclereids. These fibers can extend throughout a plant to provide support and strength to various organs. This plant tissue is commonly found in stems, bark, and in the hard shells of some fruits and nuts, such as pears. Collenchyma plant tissue is similar to sclerenchyma, in that it provides support. Often, collenchyma plant tissue is seen in young plants, with a limited number of cells. As such, only a portion of the cell wall in these cells will be thickened for support. This plant tissue is usually found wherever there is new growth and the other structural cells have not set in yet.

Complex Plant Tissue

The complex tissues in a plant deal with moving nutrients and water to the leaves, while removing the products of photosynthesis from the leaves. Photosynthesis produces the sugar glucose. Modified and bound to other 6-carbon sugars, the substance becomes sucrose or a variety of other disaccharides. In this form it can be moved with small amounts of water and can be transported efficiently throughout the plant. The complex tissues of the plant aid in this overall effort to supply the roots with food as they supply the leaves with water and nutrients.

The two main forms of plant tissue used in this process are xylem and phloem. Xylem is a plant tissue specially designed for transporting water and nutrients. This plant tissue can come in several forms, depending on the species. Sometimes, the xylem plant tissue is made up of a long chain of small tubes, called vessels, which interconnect and allow water to travel through unimpeded.

This main tube is supported by other cells, which help pull nutrients from the water and transport it to the cells within the leaves. Starting at the roots, the water is driven by pressure at the bottom and transpiration at the leaves, which sucks the water through the xylem like as straw. It is estimated that up to 95% of the water used by plants is transpired, rather than used in photosynthesis or in the metabolism. This is thought to be necessary to concentrate nutrients found in the soil.

At certain places, the xylem extends small tubes into the other type of complex plant tissue, the phloem. Like the xylem, the phloem consist of a variety of different cell types which work together to produce a continual interconnected passageway connecting cells of the plant. The phloem, rather than bringing water up from the roots, needs to carry sugar down to the roots and stems. With a little water from the xylem, it can complete this process. It is further aided by companion cells, which surround the actual sieve-tube. The whole structure is then supported by phloem fibers, which give the tube shape and structure.

Other Ways to Classify Plant Tissue

There are other ways to classify the basic plant tissue types, if the above separation seems too complicated. Some choose to classify three types of plant tissue, ground tissue, vascular tissue, and dermal tissue. This is basically the same as above, although it separates the epidermis and related tissue into the dermal category. The remaining tissues which are not vascular, it refers to as ground tissue.

Another way to classify plant tissue is based on its function. Certain tissues are only used for the purposes of photosynthesis and growth. Theses tissues can be referred to as vegetative tissue. The more specialized organs of the plant, such as flowers, fruits, and seeds, are all reproductive tissue. This method of classifying plant tissues is often used by those interested in plant genetics and reproduction, as these forms of the plant are often vastly different, genetically speaking, than the vegetative portions of the plant. Plants have a life-cycle which exhibits the alternation of generations, in which the internal portions of the flower are actually small, multicellular organisms differing genetically from the parent plant. For this reason, some scientists choose to view these tissues as separate.

Meristem

A meristem is a tissue in plants that consists of undifferentiated cells (meristematic cells) capable of cell division. Meristems give rise to various tissues and organs of a plant and are responsible for growth.

Differentiated plant cells generally cannot divide or produce cells of a different type. Meristematic cells are undifferentiated or incompletely differentiated, and are totipotent and capable of continued cell division. Division of meristematic cells provides new cells for expansion and differentiation of tissues and the initiation of new organs, providing the basic structure of the plant body. The cells are small, with no or small vacuoles and protoplasm fills the cell completely. The plastids (chloroplasts or chromoplasts), are undifferentiated, but are present in rudimentary form (proplastids). Meristematic cells are packed closely together without intercellular spaces. The cell wall is a very thin primary cell wall.

There are three types of meristematic tissues: apical (at the tips), intercalary (in the middle) and lateral (at the sides). At the meristem summit, there is a small group of slowly dividing cells, which is commonly called the central zone. Cells of this zone have a stem cell function and are essential for meristem maintenance. The proliferation and growth rates at the meristem summit usually differ considerably from those at the periphery.

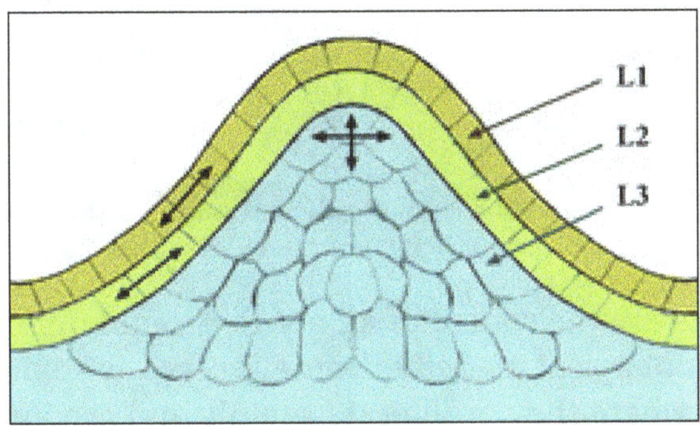

Tunica-Corpus model of the apical meristem (growing tip). The epidermal (L1) and subepidermal (L2) layers form the outer layers called the tunica. The inner L3 layer is called the corpus. Cells in the L1 and L2 layers divide in a sideways fashion, which keeps these layers distinct, whereas the L3 layer divides in a more random fashion.

Apical Meristems

Apical meristems are the completely undifferentiated (indeterminate) meristems in a plant. These differentiate into three kinds of primary meristems. The primary meristems in turn produce the two secondary meristem types. These secondary meristems are also known as lateral meristems because they are involved in lateral growth.

There are two types of apical meristem tissue: shoot apical meristem (SAM), which gives rise to organs like the leaves and flowers, and root apical meristem (RAM), which provides the meristematic cells for future root growth. SAM and RAM cells divide rapidly and are considered indeterminate, in that they do not possess any defined end status. In that sense, the meristematic cells are frequently compared to the stem cells in animals, which have an analogous behavior and function.

The number of layers varies according to plant type. In general the outermost layer is called the tunica while the innermost layers are the corpus. In monocots, the tunica determine the physical characteristics of the leaf edge and margin. In dicots, layer two of the corpus determine the characteristics of the edge of the leaf. The corpus and tunica play a critical part of the plant physical appearance as all plant cells are formed from the meristems. Apical meristems are found in two locations: the root and the stem. Some Arctic plants have an apical meristem in the lower/middle parts of the plant. It is thought that this kind of meristem evolved because it is advantageous in Arctic conditions.

Shoot Apical Meristems

Shoot apical meristems of Crassula ovata (left). Fourteen days later, leaves have developed (right).

Shoot apical meristems are the source of all above-ground organs, such as leaves and flowers. Cells at the shoot apical meristem summit serve as stem cells to the surrounding peripheral region, where they proliferate rapidly and are incorporated into differentiating leaf or flower primordia.

The shoot apical meristem is the site of most of the embryogenesis in flowering plants. Primordia of leaves, sepals, petals, stamens, and ovaries are initiated here at the rate of one every time interval, called a plastochron. It is where the first indications that flower development has been evoked are manifested. One of these indications might be the loss of apical dominance and the release of otherwise dormant cells to develop as auxiliary shoot meristems, in some species in axils of primordia as close as two or three away from the apical dome. The shoot apical meristem consists of 4 distinct cell groups:

- Stem cells.

- The immediate daughter cells of the stem cells.

- A subjacent organizing center.

- Founder cells for organ initiation in surrounding regions.

The four distinct zones mentioned above are maintained by a complex signalling pathway. In Arabidopsis thaliana, 3 interacting CLAVATA genes are required to regulate the size of the stem cell reservoir in the shoot apical meristem by controlling the rate of cell division. CLV1 and CLV2 are predicted to form a receptor complex (of the LRR receptor-like kinase family) to which CLV3 is a ligand. CLV3 shares some homology with the ESR proteins of maize, with a short 14 amino acid region being conserved between the proteins. Proteins that contain these conserved regions have been grouped into the CLE family of proteins.

CLV1 has been shown to interact with several cytoplasmic proteins that are most likely involved in downstream signalling. For example, the CLV complex has been found to be associated with Rho/Rac small GTPase-related proteins. These proteins may act as an intermediate between the CLV complex and a mitogen-activated protein kinase (MAPK), which is often involved in signalling cascades. KAPP is a kinase-associated protein phosphatase that has been shown to interact with CLV1. KAPP is thought to act as a negative regulator of CLV1 by dephosphorylating it.

Another important gene in plant meristem maintenance is WUSCHEL (shortened to WUS), which is a target of CLV signaling in addition to positively regulating CLV, thus forming a feedback loop. WUS is expressed in the cells below the stem cells of the meristem and its presence prevents the differentiation of the stem cells. CLV1 acts to promote cellular differentiation by repressing WUS activity outside of the central zone containing the stem cells.

The function of WUS in the shoot apical meristem is linked to the phytohormone cytokinin. Cytokinin activates histidine kinases which then phosphorylate histidine phosphotransfer proteins. Subsequently, the phosphate groups are transferred onto two types of Arabidopsis response regulators (ARRs): Type-B ARRS and Type-A ARRs. Type-B ARRs work as transcription factors to activate genes downstream of cytokinin, including A-ARRs. A-ARRs are similar to B-ARRs in structure; however, A-ARRs do not contain the DNA binding domains that B-ARRs have, and which are required to function as transcription factors. Therefore, A-ARRs do not contribute to the activation of transcription, and by competing for phosphates from phospho-transfer proteins, inhibit B-ARRs function. In the SAM, B-ARRs induce the expression of WUS which induces stem cell identity. WUS then suppresses A-ARRs. As a result, B-ARRs are no longer inhibited, causing sustained cytokinin signaling in the center of the shoot apical meristem. Altogether with CLAVATA signaling, this system works as a negative feedback loop. Cytokinin signaling is positively reinforced by WUS to prevent the inhibition of cytokinin signaling, while WUS promotes its own inhibitor in the form of CLV3, which ultimately keeps WUS and cytokinin signaling in check.

Root Apical Meristem

10x microscope image of root tip with meristem: In image-

- Quiescent center.

- Calyptrogen (live rootcap cells).

- Rootcap.

- Sloughed off dead rootcap cells.

- Procambium.

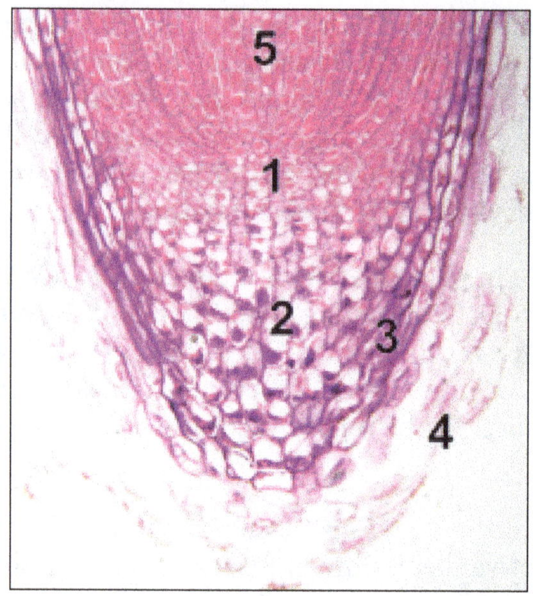

Unlike the shoot apical meristem, the root apical meristem produces cells in two dimensions. It harbors two pools of stem cells around an organizing center called the quiescent center (QC) cells and together produces most of the cells in an adult root. At its apex, the root meristem is covered by the root cap, which protects and guides its growth trajectory. Cells are continuously sloughed off the outer surface of the root cap. The QC cells are characterized by their low mitotic activity. Evidence suggests that the QC maintains the surrounding stem cells by preventing their differentiation, via signals that are yet to be discovered. This allows a constant supply of new cells in the meristem required for continuous root growth. Recent findings indicate that QC can also act as a reservoir of stem cells to replenish whatever is lost or damaged. Root apical meristem and tissue patterns become established in the embryo in the case of the primary root, and in the new lateral root primordium in the case of secondary roots.

Intercalary Meristem

In angiosperms, intercalary meristems occur only in monocot (in particular, grass) stems at the base of nodes and leaf blades. Horsetails also exhibit intercalary growth. Intercalary meristems are capable of cell division, and they allow for rapid growth and regrowth of many monocots. Intercalary meristems at the nodes of bamboo allow for rapid stem elongation, while those at the base of most grass leaf blades allow damaged leaves to rapidly regrow. This leaf regrowth in grasses evolved in response to damage by grazing herbivores.

Floral Meristem

When plants begin flowering, the shoot apical meristem is transformed into an inflorescence meristem, which goes on to produce the floral meristem, which produces the sepals, petals, stamens, and carpels of the flower.

In contrast to vegetative apical meristems and some efflorescence meristems, floral meristems cannot continue to grow indefinitely. Their future growth is limited to the flower with a particular size and form. The transition from shoot meristem to floral meristem requires floral meristem identity genes, that both specify the floral organs and cause the termination of the production of stem cells. AGAMOUS (AG) is a floral homeotic gene required for floral meristem termination and necessary for proper development of the stamens and carpels. AG is necessary to prevent the conversion of floral meristems to inflorescence shoot meristems, but is identity gene LEAFY (LFY) and WUS and is restricted to the centre of the floral meristem or the inner two whorls. This way floral identity and region specificity is achieved. WUS activates AG by binding to a consensus sequence in the AG's second intron and LFY binds to adjacent recognition sites. Once AG is activated it represses expression of WUS leading to the termination of the meristem.

Through the years, scientists have manipulated floral meristems for economic reasons. An example is the mutant tobacco plant "Maryland Mammoth". In 1936, the department of agriculture of Switzerland performed several scientific tests with this plant. "Maryland Mammoth" is peculiar in that it grows much faster than other tobacco plants.

Apical Dominance

Apical dominance is the phenomenon where one meristem prevents or inhibits the growth of other meristems. As a result, the plant will have one clearly defined main trunk. For example, in trees, the tip of the main trunk bears the dominant shoot meristem. Therefore, the tip of the trunk grows rapidly and is not shadowed by branches. If the dominant meristem is cut off, one or more branch tips will assume dominance. The branch will start growing faster and the new growth will be vertical. Over the years, the branch may begin to look more and more like an extension of the main trunk. Often several branches will exhibit this behavior after the removal of apical meristem, leading to a bushy growth.

The mechanism of apical dominance is based on auxins, types of plant growth regulators. These are produced in the apical meristem and transported towards the roots in the cambium. If apical dominance is complete, they prevent any branches from forming as long as the apical meristem is active. If the dominance is incomplete, side branches will develop.

Recent investigations into apical dominance and the control of branching have revealed a new plant hormone family termed strigolactones. These compounds were previously known to be involved in seed germination and communication with mycorrhizal fungi and are now shown to be involved in inhibition of branching.

Diversity in Meristem Architectures

The SAM contains a population of stem cells that also produce the lateral meristems while the stem elongates. It turns out that the mechanism of regulation of the stem cell number might be evolutionarily conserved. The CLAVATA gene CLV2 responsible for maintaining the stem cell population in Arabidopsis thaliana is very closely related to the Maize gene FASCIATED EAR 2(FEA2) also involved in the same function. Similarly, in Rice, the FON1-FON2 system seems to bear a close relationship with the CLV signaling system in Arabidopsis thaliana. These studies suggest that the regulation of stem cell number, identity and differentiation might be an evolutionarily conserved mechanism in monocots, if not in angiosperms. Rice also contains another genetic system distinct from FON1-FON2, that is involved in regulating stem cell number. This example underlines the innovation that goes about in the living world all the time.

Role of the KNOX-family Genes

Complex leaves of Cardamine hirsuta result from KNOX gene expression.

Genetic screens have identified genes belonging to the KNOX family in this function. These genes essentially maintain the stem cells in an undifferentiated state. The KNOX family has undergone quite a bit of evolutionary diversification while keeping the overall mechanism more or less similar. Members of the KNOX family have been found in plants as diverse as Arabidopsis thaliana, rice, barley and tomato. KNOX-like genes are also present in some algae, mosses, ferns and gymnosperms. Misexpression of these genes leads to the formation of interesting morphological features. For example, among members of Antirrhineae, only the species of the genus Antirrhinum lack a structure called spur in the floral region. A spur is considered an evolutionary innovation because it defines pollinator specificity and attraction. Researchers carried out transposon mutagenesis in Antirrhinum majus, and saw that some insertions led to formation of spurs that were very similar to the other members of Antirrhineae, indicating that the loss of spur in wild Antirrhinum majus populations could probably be an evolutionary innovation.

The KNOX family has also been implicated in leaf shape evolution. One study looked at the pattern of KNOX gene expression in A. thaliana, that has simple leaves and Cardamine hirsuta, a plant having complex leaves. In A. thaliana, the KNOX genes are completely turned off in leaves, but in C.hirsuta, the expression continued, generating complex leaves. Also, it has been proposed that the mechanism of KNOX gene action is conserved across all vascular plants, because there is a tight correlation between KNOX expression and a complex leaf morphology.

Primary Meristems

Apical meristems may differentiate into three kinds of primary meristem:

- Protoderm: Lies around the outside of the stem and develops into the epidermis.

- Procambium: Lies just inside of the protoderm and develops into primary xylem and primary phloem. It also produces the vascular cambium, and cork cambium, secondary meristems. The cork cambium further differentiates into the phelloderm (to the inside) and the phellem, or cork (to the outside). All three of these layers (cork cambium, phellem, and phelloderm) constitute the periderm. In roots, the procambium can also give rise to the pericycle, which produces lateral roots in eudicots.

- Ground meristem: Develops into the cortex and the pith. Composed of parenchyma, collenchyma and sclerenchyma cells.

These meristems are responsible for primary growth, or an increase in length or height, which were discovered by scientist Joseph D. Carr of North Carolina in 1943.

Secondary Meristems

There are two types of secondary meristems, these are also called the lateral meristems because they surround the established stem of a plant and cause it to grow laterally (i.e., larger in diameter).

- Vascular cambium, which produces secondary xylem and secondary phloem. This is a process that may continue throughout the life of the plant. This is what gives rise to wood in plants. Such plants are called arboraceous. This does not occur in plants that do not go through secondary growth (known as herbaceous plants).

- Cork cambium, which gives rise to the periderm, which replaces the epidermis.

Indeterminate Growth of Meristems

Though each plant grows according to a certain set of rules, each new root and shoot meristem can go on growing for as long as it is alive. In many plants, meristematic growth is potentially indeterminate, making the overall shape of the plant not determinate in advance. This is the primary growth. Primary growth leads to lengthening of the plant body and organ formation. All plant organs arise ultimately from cell divisions in the apical meristems, followed by cell expansion and differentiation. Primary growth gives rise to the apical part of many plants.

The growth of nitrogen-fixing root nodules on legume plants such as soybean and pea is either determinate or indeterminate. Thus, soybean (or bean and Lotus japonicus) produce determinate nodules (spherical), with a branched vascular system surrounding the central infected zone. Often, Rhizobium infected cells have only small vacuoles. In contrast, nodules on pea, clovers, and Medicago truncatula are indeterminate, to maintain (at least for some time) an active meristem that yields new cells for Rhizobium infection. Thus zones of maturity exist in the nodule. Infected cells usually possess a large vacuole. The plant vascular system is branched and peripheral.

Cloning

Under appropriate conditions, each shoot meristem can develop into a complete, new plant or clone. Such new plants can be grown from shoot cuttings that contain an apical meristem. Root apical meristems are not readily cloned, however. This cloning is called asexual reproduction or vegetative reproduction and is widely practiced in horticulture to mass-produce plants of a desirable genotype. This process is also known as mericloning.

Propagating through cuttings is another form of vegetative propagation that initiates root or shoot production from secondary meristematic cambial cells. This explains why basal 'wounding' of shoot-borne cuttings often aids root formation.

Induced Meristems

Meristems may also be induced in the roots of legumes such as soybean, Lotus japonicus, pea, and Medicago truncatula after infection with soil bacteria commonly called Rhizobia. Cells of the inner or outer cortex in the so-called "window of nodulation" just behind the developing root tip are induced to divide. The critical signal substance is the lipo-oligosaccharide Nod factor, decorated with side groups to allow specificity of interaction. The Nod factor receptor proteins NFR1 and NFR5 were cloned from several legumes including Lotus japonicus, Medicago truncatula and soybean (Glycine max). Regulation of nodule meristems utilizes long-distance regulation known as the autoregulation of nodulation (AON). This process involves a leaf-vascular tissue located LRR receptor kinases (LjHAR1, GmNARK and MtSUNN), CLE peptide signalling, and KAPP interaction, similar to that seen in the CLV1,2,3 system. LjKLAVIER also exhibits a nodule regulation phenotype though it is not yet known how this relates to the other AON receptor kinases.

Vascular Tissue

Vascular tissue is a complex conducting tissue, formed of more than one cell type, found in vascular plants. The primary components of vascular tissue are the xylem and phloem. These two tissues transport fluid and nutrients internally. There are also two meristems associated with vascular tissue: the vascular cambium and the cork cambium. All the vascular tissues within a particular plant together constitute the vascular tissue system of that plant.

Cross section of celery stalk, showing vascular bundles, which include both phloem and xylem.

Detail of the vasculature of a bramble leaf.

The cells in vascular tissue are typically long and slender. Since the xylem and phloem function in the conduction of water, minerals, and nutrients throughout the plant, it is not surprising that their form should be similar to pipes. The individual cells of phloem are connected end-to-end, just as the sections of a pipe might be. As the plant grows, new vascular tissue differentiates in the growing tips of the plant. The new tissue is aligned with existing vascular tissue, maintaining its connection throughout the plant. The vascular tissue in plants is arranged in long, discrete strands called vascular bundles. These bundles include both xylem and phloem, as well as supporting and protective cells. In stems and roots, the xylem typically lies closer to the interior of the stem with phloem towards the exterior of the stem. In the stems of some Asterales dicots, there may be phloem located inwardly from the xylem as well.

Between the xylem and phloem is a meristem called the vascular cambium. This tissue divides off cells that will become additional xylem and phloem. This growth increases the girth of the plant, rather than its length. As long as the vascular cambium continues to produce new cells, the plant will continue to grow more stout. In trees and other plants that develop wood, the vascular cambium allows the expansion of vascular tissue that produces woody growth. Because this growth ruptures the epidermis of the stem, woody plants also have a cork cambium that develops among the phloem. The cork cambium gives rise to thickened cork cells to protect the surface of the plant and reduce water loss. Both the production of wood and the production of cork are forms of secondary growth.

In leaves, the vascular bundles are located among the spongy mesophyll. The xylem is oriented toward the adaxial surface of the leaf (usually the upper side), and phloem is oriented toward the abaxial surface of the leaf. This is why aphids are typically found on the undersides of the leaves rather than on the top, since the phloem transports sugars manufactured by the plant and they are closer to the lower surface.

Xylem

Xylem (blue) transports water and minerals from the roots upwards.

Xylem is one of the two types of transport tissue in vascular plants, phloem being the other. The basic function of xylem is to transport water from roots to stems and leaves, but it also transports nutrients.

Structure

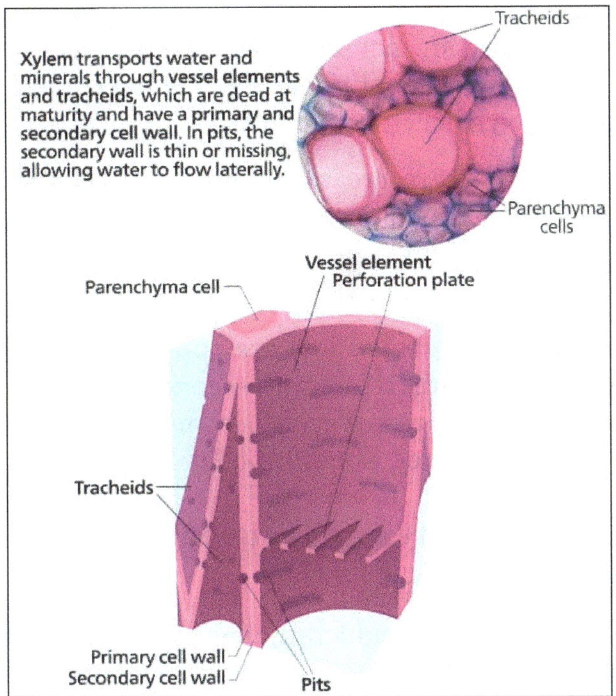

Cross section of some xylem cells.

The most distinctive xylem cells are the long tracheary elements that transport water. Tracheids and vessel elements are distinguished by their shape; vessel elements are shorter, and are connected together into long tubes that are called vessels.

Xylem also contains two other cell types: parenchyma and fibers.

Xylem can be found:

- In vascular bundles, present in non-woody plants and non-woody parts of woody plants.

- In secondary xylem, laid down by a meristem called the vascular cambium in woody plants.

- As part of a stelar arrangement not divided into bundles, as in many ferns.

In transitional stages of plants with secondary growth, the first two categories are not mutually exclusive, although usually a vascular bundle will contain primary xylem only.

The branching pattern exhibited by xylem follows Murray's law.

Primary and Secondary Xylem

Primary xylem is formed during primary growth from procambium. It includes protoxylem and metaxylem. Metaxylem develops after the protoxylem but before secondary xylem. Metaxylem has wider vessels and tracheids than protoxylem.

Secondary xylem is formed during secondary growth from vascular cambium. Although secondary xylem is also found in members of the gymnosperm groups Gnetophyta and Ginkgophyta and to a lesser extent in members of the Cycadophyta, the two main groups in which secondary xylem can be found are:

- Conifers (Coniferae): There are approximately 600 known species of conifers. All species have secondary xylem, which is relatively uniform in structure throughout this group. Many conifers become tall trees: the secondary xylem of such trees is used and marketed as softwood.

- Angiosperms (Angiospermae): There are approximately 250,000 known species of angiosperms. Within this group secondary xylem is rare in the monocots. Many non-monocot angiosperms become trees, and the secondary xylem of these is used and marketed as hardwood.

Main Function – Upwards Water Transport

The xylem, vessels and tracheids of the roots, stems and leaves are interconnected to form a continuous system of water-conducting channels reaching all parts of the plants. The system transports water and soluble mineral nutrients from the roots throughout the plant. It is also used to replace water lost during transpiration and photosynthesis. Xylem sap consists mainly of water and inorganic ions, although it can also contain a number of organic chemicals as well. The transport is passive, not powered by energy spent by the tracheary elements themselves, which are dead by maturity and no longer have living contents. Transporting sap upwards becomes more difficult as the height of a plant increases and upwards transport of water by xylem is considered to limit the maximum height of trees. Three phenomena cause xylem sap to flow:

- Pressure flow hypothesis: Sugars produced in the leaves and other green tissues are kept in the phloem system, creating a solute pressure differential versus the xylem system carrying a far lower load of solutes- water and minerals. The phloem pressure can rise to several MPa, far higher than atmospheric pressure. Selective inter-connection between these systems allows this high solute concentration in the phloem to draw xylem fluid upwards by negative pressure.

- Transpirational pull: Similarly, the evaporation of water from the surfaces of mesophyll cells to the atmosphere also creates a negative pressure at the top of a plant. This causes millions of minute menisci to form in the mesophyll cell wall. The resulting surface tension causes a negative pressure or tension in the xylem that pulls the water from the roots and soil.

- Root pressure: If the water potential of the root cells is more negative than that of the soil, usually due to high concentrations of solute, water can move by osmosis into the root from the soil. This causes a positive pressure that forces sap up the xylem towards the leaves. In some circumstances, the sap will be forced from the leaf through a hydathode in a phenomenon known as guttation. Root pressure is highest in the morning before the stomata open and allow transpiration to begin. Different plant species can have different root pressures even in a similar environment; examples include up to 145 kPa in Vitis riparia but around zero in Celastrus orbiculatus.

The primary force that creates the capillary action movement of water upwards in plants is the adhesion between the water and the surface of the xylem conduits. Capillary action provides the force that establishes an equilibrium configuration, balancing gravity. When transpiration removes water at the top, the flow is needed to return to the equilibrium.

Transpirational pull results from the evaporation of water from the surfaces of cells in the leaves. This evaporation causes the surface of the water to recess into the pores of the cell wall. By capillary action, the water forms concave menisci inside the pores. The high surface tension of water pulls the concavity outwards, generating enough force to lift water as high as a hundred meters from ground level to a tree's highest branches.

Transpirational pull requires that the vessels transporting the water be very small in diameter; otherwise, cavitation would break the water column. And as water evaporates from leaves, more is drawn up through the plant to replace it. When the water pressure within the xylem reaches extreme levels due to low water input from the roots (if, for example, the soil is dry), then the gases come out of solution and form a bubble – an embolism forms, which will spread quickly to other adjacent cells, unless bordered pits are present (these have a plug-like structure called a torus, that seals off the opening between adjacent cells and stops the embolism from spreading).

Cohesion-tension Theory

The cohesion-tension theory is a theory of intermolecular attraction that explains the process of water flow upwards (against the force of gravity) through the xylem of plants. It was proposed in 1894 by John Joly and Henry Horatio Dixon. Despite numerous objections, this is the most widely accepted theory for the transport of water through a plant's vascular system based on the classical research of Dixon-Joly, Eugen Askenasy and Dixon.

Water is a polar molecule. When two water molecules approach one another, the slightly negatively charged oxygen atom of one forms a hydrogen bond with a slightly positively charged hydrogen atom in the other. This attractive force, along with other intermolecular forces, is one of the principal factors responsible for the occurrence of surface tension in liquid water. It also allows plants to draw water from the root through the xylem to the leaf.

Water is constantly lost through transpiration from the leaf. When one water molecule is lost another is pulled along by the processes of cohesion and tension. Transpiration pull, utilizing capillary action and the inherent surface tension of water, is the primary mechanism of water movement in plants. However, it is not the only mechanism involved. Any use of water in leaves forces water to move into them.

Transpiration in leaves creates tension (differential pressure) in the cell walls of mesophyll cells. Because of this tension, water is being pulled up from the roots into the leaves, helped by cohesion (the pull between individual water molecules, due to hydrogen bonds) and adhesion (the stickiness between water molecules and the hydrophilic cell walls of plants). This mechanism of water flow works because of water potential (water flows from high to low potential), and the rules of simple diffusion.

Over the past century, there has been a great deal of research regarding the mechanism of xylem sap transport; today, most plant scientists continue to agree that the cohesion-tension theory best explains this process, but multiforce theories that hypothesize several alternative mechanisms have been suggested, including longitudinal cellular and xylem osmotic pressure gradients, axial potential gradients in the vessels, and gel- and gas-bubble-supported interfacial gradients.

Measurement of Pressure

Until recently, the differential pressure (suction) of transpirational pull could only be measured indirectly, by applying external pressure with a pressure bomb to counteract it. When the technology to perform direct measurements with a pressure probe was developed, there was initially some doubt about whether the classic theory was correct, because some workers were unable to demonstrate negative pressures. More recent measurements do tend to validate the classic theory, for the most part. Xylem transport is driven by a combination of transpirational pull from above and root pressure from below, which makes the interpretation of measurements more complicated.

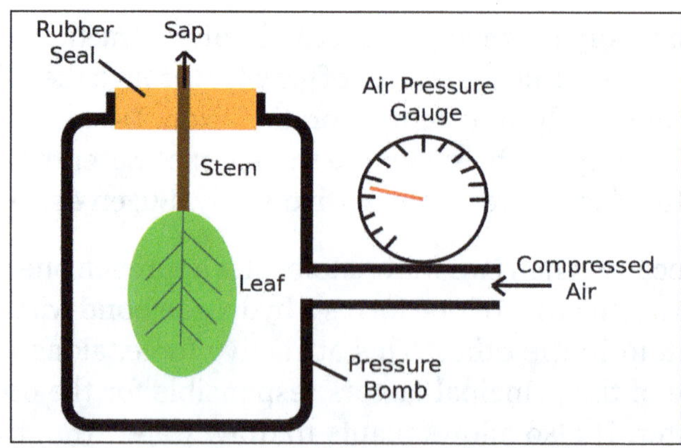

A diagram showing the setup of a pressure bomb.

Evolution

Xylem appeared early in the history of terrestrial plant life. Fossil plants with anatomically preserved xylem are known from the Silurian (more than 400 million years ago), and trace fossils resembling individual xylem cells may be found in earlier Ordovician rocks. The earliest true and recognizable xylem consists of tracheids with a helical-annular reinforcing layer added to the cell wall. This is the only type of xylem found in the earliest vascular plants, and this type of cell continues to be found in the protoxylem (first-formed xylem) of all living groups of vascular plants. Several groups of plants later developed pitted tracheid cells independently through convergent evolution. In living plants, pitted tracheids do not appear in development until the maturation of the metaxylem (following the protoxylem).

In most plants, pitted tracheids function as the primary transport cells. The other type of vascular element, found in angiosperms, is the vessel element. Vessel elements are joined end to end to form vessels in which water flows unimpeded, as in a pipe. The presence of xylem vessels is considered to be one of the key innovations that led to the success of the angiosperms. However, the occurrence of vessel elements is not restricted to angiosperms, and they are absent in some archaic or "basal" lineages of the angiosperms: (e.g., Amborellaceae, Tetracentraceae, Trochodendraceae, and Winteraceae), and their secondary xylem is described by Arthur Cronquist as "primitively vesselless". Cronquist considered the vessels of Gnetum to be convergent with those of angiosperms. Whether the absence of vessels in basal angiosperms is a primitive condition is contested, the alternative hypothesis states that vessel elements originated in a precursor to the angiosperms and were subsequently lost.

To photosynthesize, plants must absorb CO_2 from the atmosphere. However, this comes at a price: while stomata are open to allow CO_2 to enter, water can evaporate. Water is lost much faster than CO_2 is absorbed, so plants need to replace it, and have developed systems to transport water from the moist soil to the site of photosynthesis. Early plants sucked water between the walls of their cells, then evolved the ability to control water loss (and CO_2 acquisition) through the use of stomata. Specialized water transport tissues soon evolved in the form of hydroids, tracheids, then secondary xylem, followed by an endodermis and ultimately vessels.

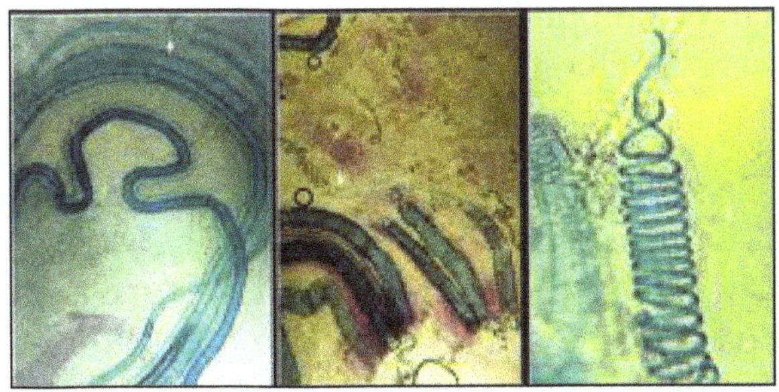

Photos showing xylem elements in the shoot of a fig tree (Ficus alba): crushed in hydrochloric acid, between slides and cover slips.

The high CO_2 levels of Silurian-Devonian times, when plants were first colonizing land, meant that the need for water was relatively low. As CO_2 was withdrawn from the atmosphere by plants, more water was lost in its capture, and more elegant transport mechanisms evolved. As water transport mechanisms, and waterproof cuticles, evolved, plants could survive without being continually covered by a film of water. This transition from poikilohydry to homoiohydry opened up new potential for colonization. Plants then needed a robust internal structure that held long narrow channels for transporting water from the soil to all the different parts of the above-soil plant, especially to the parts where photosynthesis occurred.

During the Silurian, CO_2 was readily available, so little water needed expending to acquire it. By the end of the Carboniferous, when CO_2 levels had lowered to something approaching today's, around 17 times more water was lost per unit of CO_2 uptake. However, even in these "easy" early days, water was at a premium, and had to be transported to parts of the plant from the wet soil to avoid desiccation. This early water transport took advantage of the cohesion-tension mechanism inherent in water. Water has a tendency to diffuse to areas that are drier, and this process is accelerated when water can be wicked along a fabric with small spaces. In small passages, such as that between the plant cell walls (or in tracheids), a column of water behaves like rubber – when molecules evaporate from one end, they pull the molecules behind them along the channels. Therefore, transpiration alone provided the driving force for water transport in early plants. However, without dedicated transport vessels, the cohesion-tension mechanism cannot transport water more than about 2 cm, severely limiting the size of the earliest plants. This process demands a steady supply of water from one end, to maintain the chains; to avoid exhausting it, plants developed a waterproof cuticle. Early cuticle may not have had pores but did not cover the entire plant surface, so that gas exchange could continue. However, dehydration at times was inevitable; early plants cope with this by having a lot of water stored between their cell walls, and when it comes to it sticking out the tough times by putting life "on hold" until more water is supplied.

To be free from the constraints of small size and constant moisture that the parenchymatic transport system inflicted, plants needed a more efficient water transport system. During the early Silurian, they developed specialized cells, which were lignified (or bore similar chemical compounds) to avoid implosion; this process coincided with cell death, allowing their innards to be emptied and water to be passed through them. These wider, dead, empty cells were a million times more conductive than the inter-cell method, giving the potential for transport over longer distances, and higher CO_2 diffusion rates.

The earliest macrofossils to bear water-transport tubes are Silurian plants placed in the genus Cooksonia. The early Devonian pretracheophytes Aglaophyton and Horneophyton have structures very similar to the hydroids of modern mosses. Plants continued to innovate new ways of reducing the resistance to flow within their cells, thereby increasing the efficiency of their water transport. Bands on the walls of tubes, in fact apparent from the early Silurian onwards, are an early improvisation to aid the easy flow of water. Banded tubes, as well as tubes with pitted ornamentation on their walls, were lignified and, when they form single celled conduits, are considered to be tracheids. These, the "next generation" of transport cell design, have a more rigid structure than hydroids, allowing them to cope with higher levels of water pressure. Tracheids may have a single evolutionary origin, possibly within the hornworts, uniting all tracheophytes (but they may have evolved more than once).

Water transport requires regulation, and dynamic control is provided by stomata. By adjusting the amount of gas exchange, they can restrict the amount of water lost through transpiration. This is an important role where water supply is not constant, and indeed stomata appear to have evolved before tracheids, being present in the non-vascular hornworts.

An endodermis probably evolved during the Silu-Devonian, but the first fossil evidence for such a structure is Carboniferous. This structure in the roots covers the water transport tissue and regulates ion exchange (and prevents unwanted pathogens etc. from entering the water transport system). The endodermis can also provide an upwards pressure, forcing water out of the roots when transpiration is not enough of a driver.

Once plants had evolved this level of controlled water transport, they were truly homoiohydric, able to extract water from their environment through root-like organs rather than relying on a film of surface moisture, enabling them to grow to much greater size. As a result of their independence from their surroundings, they lost their ability to survive desiccation – a costly trait to retain.

During the Devonian, maximum xylem diameter increased with time, with the minimum diameter remaining pretty constant. By the middle Devonian, the tracheid diameter of some plant lineages (Zosterophyllophytes) had plateaued. Wider tracheids allow water to be transported faster, but the overall transport rate depends also on the overall cross-sectional area of the xylem bundle itself. The increase in vascular bundle thickness further seems to correlate with the width of plant axes, and plant height; it is also closely related to the appearance of leaves and increased stomatal density, both of which would increase the demand for water.

While wider tracheids with robust walls make it possible to achieve higher water transport pressures, this increases the problem of cavitation. Cavitation occurs when a bubble of air forms within a vessel, breaking the bonds between chains of water molecules and preventing them from pulling more water up with their cohesive tension. A tracheid, once cavitated, cannot have its embolism removed and return to service (except in a few advanced angiosperms which have developed a mechanism of doing so). Therefore, it is well worth plants' while to avoid cavitation occurring. For this reason, pits in tracheid walls have very small diameters, to prevent air entering and allowing bubbles to nucleate. Freeze-thaw cycles are a major cause of cavitation. Damage to a tracheid's wall almost inevitably leads to air leaking in and cavitation, hence the importance of many tracheids working in parallel.

Cavitation is hard to avoid, but once it has occurred plants have a range of mechanisms to contain the damage. Small pits link adjacent conduits to allow fluid to flow between them, but not air – although ironically these pits, which prevent the spread of embolisms, are also a major cause of them. These pitted surfaces further reduce the flow of water through the xylem by as much as 30%. Conifers, by the Jurassic, developed an ingenious improvement, using valve-like structures to isolate cavitated elements. These torus-margo structures have a blob floating in the middle of a donut; when one side depressurizes the blob is sucked into the torus and blocks further flow. Other plants simply accept cavitation; for instance, oaks grow a ring of wide vessels at the start of each spring, none of which survive the winter frosts. Maples use root pressure each spring to force sap upwards from the roots, squeezing out any air bubbles.

Growing to height also employed another trait of tracheids – the support offered by their lignified walls. Defunct tracheids were retained to form a strong, woody stem, produced in most instances by a secondary xylem. However, in early plants, tracheids were too mechanically vulnerable, and retained a central position, with a layer of tough sclerenchyma on the outer rim of the stems. Even when tracheids do take a structural role, they are supported by sclerenchymatic tissue.

Tracheids end with walls, which impose a great deal of resistance on flow; vessel members have perforated end walls, and are arranged in series to operate as if they were one continuous vessel. The function of end walls, which were the default state in the Devonian, was probably to avoid embolisms. An embolism is where an air bubble is created in a tracheid. This may happen as a result of freezing, or by gases dissolving out of solution. Once an embolism is formed, it usually cannot be removed; the affected cell cannot pull water up, and is rendered useless.

End walls excluded, the tracheids of prevascular plants were able to operate under the same hydraulic conductivity as those of the first vascular plant, Cooksonia.

The size of tracheids is limited as they comprise a single cell; this limits their length, which in turn limits their maximum useful diameter to 80 µm. Conductivity grows with the fourth power of diameter, so increased diameter has huge rewards; vessel elements, consisting of a number of cells, joined at their ends, overcame this limit and allowed larger tubes to form, reaching diameters of up to 500 µm, and lengths of up to 10 m.

Vessels first evolved during the dry, low CO_2 periods of the late Permian, in the horsetails, ferns and Selaginellales independently, and later appeared in the mid Cretaceous in angiosperms and gnetophytes. Vessels allow the same cross-sectional area of wood to transport around a hundred times more water than tracheids. This allowed plants to fill more of their stems with structural fibers, and also opened a new niche to vines, which could transport water without being as thick as the tree they grew on. Despite these advantages, tracheid-based wood is a lot lighter, thus cheaper to make, as vessels need to be much more reinforced to avoid cavitation.

Development

Xylem development can be described by four terms: centrarch, exarch, endarch and mesarch. As it develops in young plants, its nature changes from protoxylem to metaxylem (i.e. from first xylem to after xylem). The patterns in which protoxylem and metaxylem are arranged is important in the study of plant morphology.

Protoxylem and Metaxylem

As a young vascular plant grows, one or more strands of primary xylem form in its stems and roots. The first xylem to develop is called 'protoxylem'. In appearance protoxylem is usually distinguished by narrower vessels formed of smaller cells. Some of these cells have walls which contain thickenings in the form of rings or helices. Functionally, protoxylem can extend: the cells are able to grow in size and develop while a stem or root is elongating. Later, 'metaxylem' develops in the strands of xylem. Metaxylem vessels and cells are usually larger; the cells have thickenings which are typically either in the form of ladderlike transverse bars (scalariform) or continuous sheets except for holes or pits (pitted). Functionally, metaxylem completes its development after elongation ceases when the cells no longer need to grow in size.

Patterns of Protoxylem and Metaxylem

There are four main patterns to the arrangement of protoxylem and metaxylem in stems and roots.

- Centrarch refers to the case in which the primary xylem forms a single cylinder in the center of the stem and develops from the center outwards. The protoxylem is thus found in the central core and the metaxylem in a cylinder around it. This pattern was common in early land plants, such as "rhyniophytes", but is not present in any living plants.

- Exarch is used when there is more than one strand of primary xylem in a stem or root, and the xylem develops from the outside inwards towards the center, i.e. centripetally. The metaxylem is thus closest to the center of the stem or root and the protoxylem closest to the periphery. The roots of vascular plants are normally considered to have exarch development.

- Endarch is used when there is more than one strand of primary xylem in a stem or root, and the xylem develops from the inside outwards towards the periphery, i.e. centrifugally. The protoxylem is thus closest to the center of the stem or root and the metaxylem closest to the periphery. The stems of seed plants typically have endarch development.

- Mesarch is used when there is more than one strand of primary xylem in a stem or root, and the xylem develops from the middle of a strand in both directions. The metaxylem is thus on both the peripheral and central sides of the strand with the protoxylem between the metaxylem (possibly surrounded by it). The leaves and stems of many ferns have mesarch development.

Phloem

Phloem is the living tissue in vascular plants that transports the soluble organic compounds made during photosynthesis and known as photosynthates, in particular the sugar sucrose, to parts of the plant where needed. This transport process is called translocation. The term was introduced by Carl Nägeli in 1858.

Structure

Phloem tissue consists of conducting cells, generally called sieve elements, parenchyma cells, including both specialized companion cells or albuminous cells and unspecialized cells and supportive cells, such as fibres and sclereids.

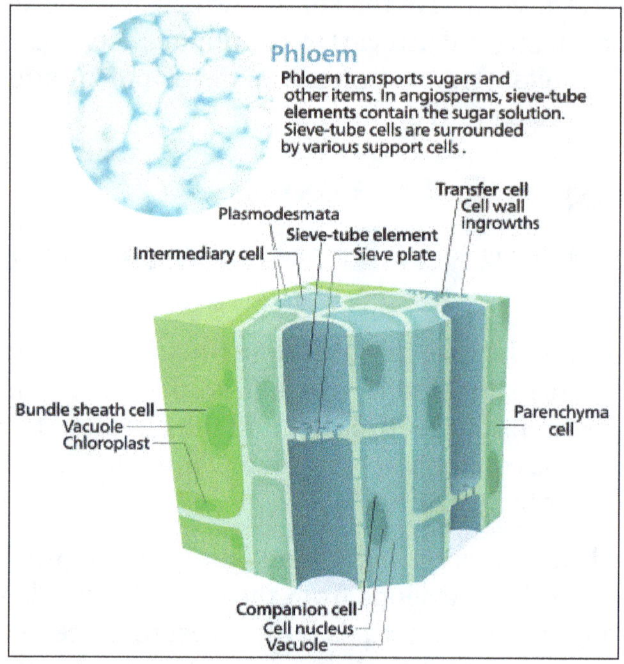

Cross section of some phloem cells.

Conducting Cells (Sieve Elements)

Sieve elements are the type of cell that are responsible for transporting sugars throughout the plant. At maturity they lack a nucleus and have very few organelles, so they rely on companion cells or albuminous cells for most of their metabolic needs. Sieve tube cells do contain vacuoles and other organelles, such as ribosomes, before they mature, but these generally migrate to the cell wall and dissolve at maturity; this ensures there is little to impede the movement of fluids. One of the few organelles they do contain at maturity is the rough endoplasmic reticulum, which can be found at the plasma membrane, often nearby the plasmodesmata that connect them to their companion or albuminous cells. All sieve cells have groups of pores at their ends that grow from modified and enlarged plasmodesmata, called sieve areas. The pores are reinforced by platelets of a polysaccharide called callose.

Parenchyma Cells

They are of two types, aerenchyma and chlorenchyma. Other parenchyma cells within the phloem are generally undifferentiated and used for food storage.

Companion Cells

The metabolic functioning of sieve-tube members depends on a close association with the companion cells, a specialized form of parenchyma cell. All of the cellular functions of a sieve-tube element are carried out by the (much smaller) companion cell, a typical nucleate plant cell except the companion cell usually has a larger number of ribosomes and mitochondria. The dense cytoplasm of a companion cell is connected to the sieve-tube element by plasmodesmata. The common sidewall shared by a sieve tube element and a companion cell has large numbers of plasmodesmata.

There are two types of companion cells.

- Ordinary companion cells, which have smooth walls and few or no plasmodesmatal connections to cells other than the sieve tube.

- Transfer cells, which have much-folded walls that are adjacent to non-sieve cells, allowing for larger areas of transfer. They are specialized in scavenging solutes from those in the cell walls that are actively pumped requiring energy.

Albuminous Cells

Albuminous cells have a similar role to companion cells, but are associated with sieve cells only and are hence found only in seedless vascular plants and gymnosperms.

Supportive Cells

Although its primary function is transport of sugars, phloem may also contain cells that have a mechanical support function. These generally fall into two categories: fibres and sclereids. Both cell types have a secondary cell wall and are therefore dead at maturity. The secondary cell wall increases their rigidity and tensile strength.

Fibres

Bast fibres are the long, narrow supportive cells that provide tension strength without limiting flexibility. They are also found in xylem, and are the main component of many textiles such as paper, linen, and cotton.

Sclereids

Sclereids are irregularly shaped cells that add compression strength but may reduce flexibility to some extent. They also serve as anti-herbivory structures, as their irregular shape and hardness will increase wear on teeth as the herbivores chews. For example, they are responsible for the gritty texture in pears, and in winter bears.

Function

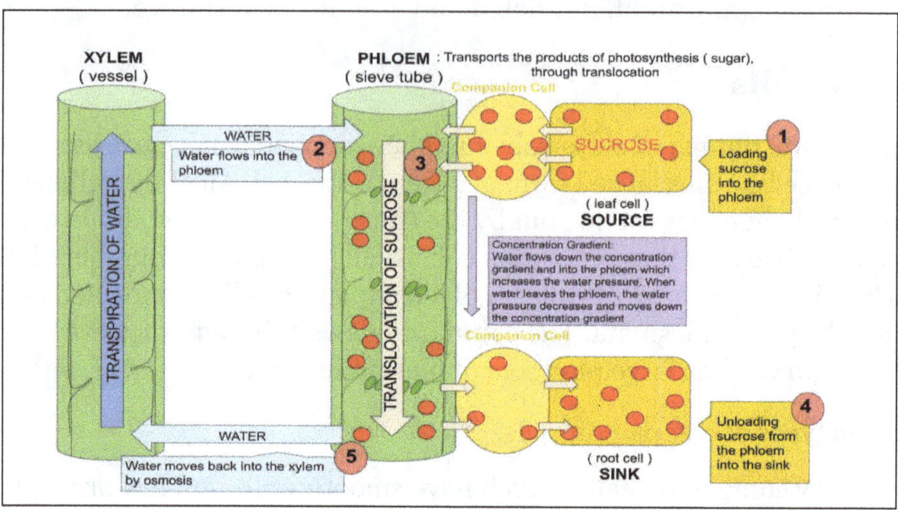

Unlike xylem (which is composed primarily of dead cells), the phloem is composed of still-living cells that transport sap. The sap is a water-based solution, but rich in sugars made by photosynthesis. These sugars are transported to non-photosynthetic parts of the plant, such as the roots, or into storage structures, such as tubers or bulbs.

During the plant's growth period, usually during the spring, storage organs such as the roots are sugar sources, and the plant's many growing areas are sugar sinks. The movement in phloem is multidirectional, whereas, in xylem cells, it is unidirectional (upward).

After the growth period, when the meristems are dormant, the leaves are sources, and storage organs are sinks. Developing seed-bearing organs (such as fruit) are always sinks. Because of this multi-directional flow, coupled with the fact that sap cannot move with ease between adjacent sieve-tubes, it is not unusual for sap in adjacent sieve-tubes to be flowing in opposite directions.

While movement of water and minerals through the xylem is driven by negative pressures (tension) most of the time, movement through the phloem is driven by positive hydrostatic pressures. This process is termed translocation, and is accomplished by a process called phloem loading and unloading.

Phloem sap is also thought to play a role in sending informational signals throughout vascular plants. "Loading and unloading patterns are largely determined by the conductivity and number of plasmodesmata and the position-dependent function of solute-specific, plasma membrane transport proteins. Recent evidence indicates that mobile proteins and RNA are part of the plant's long-distance communication signaling system. Evidence also exists for the directed transport and sorting of macromolecules as they pass through plasmodesmata".

Organic molecules such as sugars, amino acids, certain hormones, and even messenger RNAs are transported in the phloem through sieve tube elements.

Phloem is also used as a popular site for oviposition and breeding of insects belonging to the order Diptera, including the fruit fly Drosophila montana.

Girdling

Because phloem tubes are located outside the xylem in most plants, a tree or other plant can be killed by stripping away the bark in a ring on the trunk or stem. With the phloem destroyed, nutrients cannot reach the roots, and the tree/plant will die. Trees located in areas with animals such as beavers are vulnerable since beavers chew off the bark at a fairly precise height. This process is known as girdling, and can be used for agricultural purposes. For example, enormous fruits and vegetables seen at fairs and carnivals are produced via girdling. A farmer would place a girdle at the base of a large branch, and remove all but one fruit/vegetable from that branch. Thus, all the sugars manufactured by leaves on that branch have no sinks to go to but the one fruit/vegetable, which thus expands to many times its normal size.

Origin

When the plant is an embryo, vascular tissue emerges from procambium tissue, which is at the center of the embryo. Protophloem itself appears in the mid-vein extending into the cotyledonary node, which constitutes the first appearance of a leaf in angiosperms, where it forms continuous strands. The hormone auxin, transported by the protein PIN1 is responsible for the growth of those protophloem strands, signaling the final identity of those tissues. SHORTROOT(SHR), and microRNA165/166 also participate in that process, while Callose Synthase 3(CALS3), inhibits the locations where SHORTROOT(SHR), and microRNA165 can go.

In the embryo, root phloem develops independently in the upper hypocotyl, which lies between the embryonic root, and the cotyledon.

In an adult, the phloem originates, and grows outwards from, meristematic cells in the vascular cambium. Phloem is produced in phases. Primary phloem is laid down by the apical meristem and develops from the procambium. Secondary phloem is laid down by the vascular cambium to the inside of the established layer(s) of phloem.

In some eudicot families (Apocynaceae, Convolvulaceae, Cucurbitaceae, Solanaceae, Myrtaceae, Asteraceae, Thymelaeaceae), phloem also develops on the inner side of the vascular cambium; in this case, a distinction between external and internal or intraxylary phloem is made. Internal phloem is mostly primary, and begins differentiation later than the external phloem and protoxylem, though it is not without exceptions. In some other families (Amaranthaceae, Nyctaginaceae, Salvadoraceae), the cambium also periodically forms inward strands or layers of phloem, embedded in the xylem. Such phloem strands are called included or interxylary phloem.

Epidermis

The epidermis is a single layer of cells that covers the leaves, flowers, roots and stems of plants. It forms a boundary between the plant and the external environment. The epidermis serves several functions: it protects against water loss, regulates gas exchange, secretes metabolic compounds, and (especially in roots) absorbs water and mineral nutrients. The epidermis of most leaves shows dorsoventral anatomy: the upper (adaxial) and lower (abaxial) surfaces have somewhat different construction and may serve different functions. Woody stems and some other stem structures such as potato tubers produce a secondary covering called the periderm that replaces the epidermis as the protective covering.

The epidermis is the outermost cell layer of the primary plant body. In some older works the cells of the leaf epidermis have been regarded as specialised parenchyma cells, but the established modern preference has long been to classify the epidermis as dermal tissue, whereas parenchyma is classified as ground tissue. The epidermis is the main component of the dermal tissue system of leaves (diagrammed below), and also stems, roots, flowers, fruits, and seeds; it is usually transparent (epidermal cells have fewer chloroplasts or lack them completely, except for the guard cells).

The cells of the epidermis are structurally and functionally variable. Most plants have an epidermis that is a single cell layer thick. Some plants like Ficus elastica and Peperomia, which have periclinal cellular division within the protoderm of the leaves, have an epidermis with multiple cell layers. Epidermal cells are tightly linked to each other and provide mechanical strength and protection to the plant. The walls of the epidermal cells of the above ground parts of plants contain cutin, and are covered with a cuticle. The cuticle reduces water loss to the atmosphere, it is sometimes covered with wax in smooth sheets, granules, plates, tubes or filaments. The wax layers give some plants a whitish or bluish surface color. Surface wax acts as a moisture barrier and protects the plant from intense sunlight and wind. The underside of many leaves have a thinner cuticle than the top side, and leaves of plants from dry climates often have thickened cuticles to conserve water by reducing transpiration.

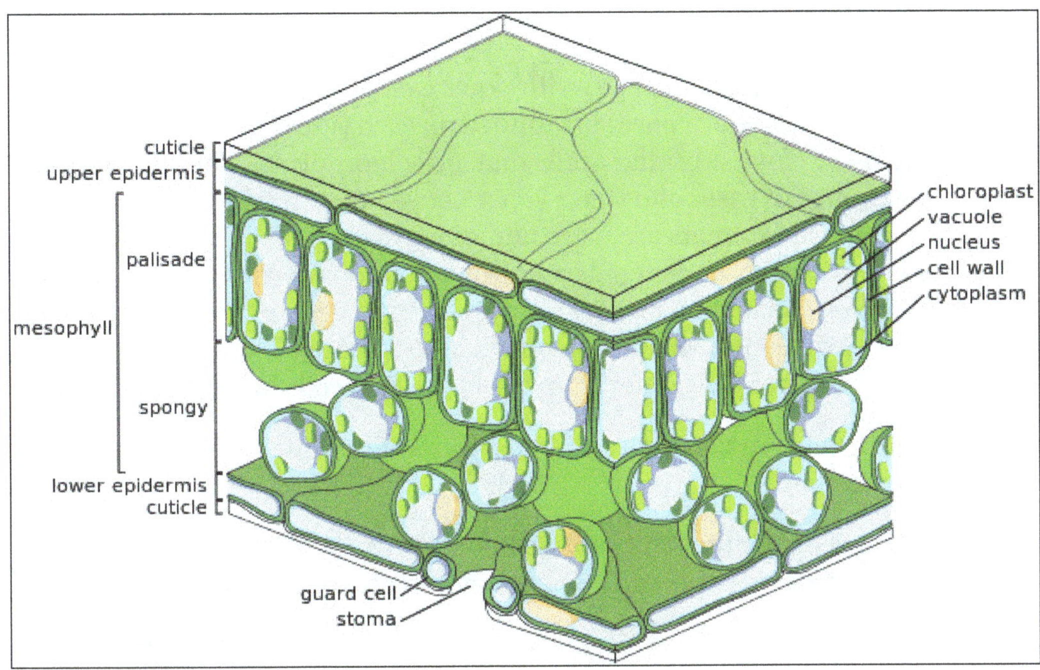

The epidermal tissue includes several differentiated cell types: epidermal cells, guard cells, subsidiary cells, and epidermal hairs (trichomes). The epidermal cells are the most numerous, largest, and least specialized. These are typically more elongated in the leaves of monocots than in those of dicots.

Trichomes or hairs grow out from the epidermis in many species. In root epidermis, epidermal hairs, termed root hairs are common and are specialized for absorption of water and mineral nutrients.

In plants with secondary growth, the epidermis of roots and stems is usually replaced by a periderm through the action of a cork cambium.

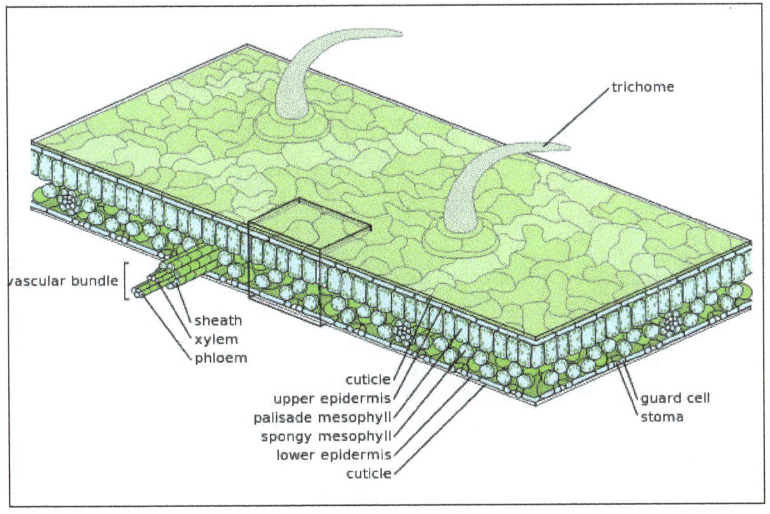

Guard Cells

The leaf and stem epidermis is covered with pores called stomata (sing., stoma), part of a stoma complex consisting of a pore surrounded on each side by chloroplast-containing guard cells, and two to four subsidiary cells that lack chloroplasts. The stomata complex regulates the exchange of gases and water vapor between the outside air and the interior of the leaf. Typically, the stomata are more numerous over the abaxial (lower) epidermis of the leaf than the (adaxial) upper epidermis. An exception is floating leaves where most or all stomata are on the upper surface. Vertical leaves, such as those of many grasses, often have roughly equal numbers of stomata on both surfaces. The stoma is bounded by two guard cells. The guard cells differ from the epidermal cells in the following aspects:

- The guard cells are bean-shaped in surface view, while the epidermal cells are irregular in shape

- The guard cells contain chloroplasts, so they can manufacture food by photosynthesis (The epidermal cells do not contain chloroplasts).

- Guard cells are the only epidermal cells that can make sugar. According to one theory, in sunlight the concentration of potassium ions (K+) increases in the guard cells. This, together with the sugars formed, lowers the water potential in the guard cells. As a result, water from other cells enter the guard cells by osmosis so they swell and become turgid. Because the guard cells have a thicker cellulose wall on one side of the cell, i.e. the side around the stomatal pore, the swollen guard cells become curved and pull the stomata open.

At night, the sugar is used up and water leaves the guard cells, so they become flaccid and the stomatal pore closes. In this way, they reduce the amount of water vapour escaping from the leaf.

Cell Differentiation in the Epidermis

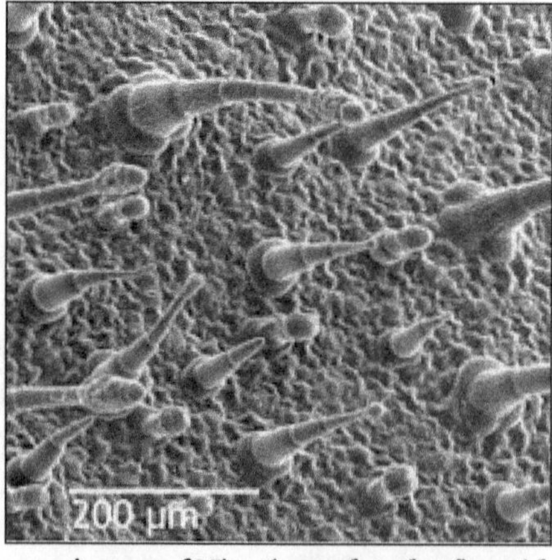

Scanning electron microscope image of Nicotiana alata leaf's epidermis, showing trichomes (hair-like appendages) and stomata (eye-shaped slits, visible at full resolution).

The plant epidermis consists of three main cell types: pavement cells, guard cells and their subsidiary cells that surround the stomata and trichomes, otherwise known as leaf hairs. The epidermis of petals also form a variation of trichomes called conical cells.

Trichomes develop at a distinct phase during leaf development, under the control of two major trichome specification genes: TTG and GL1. The process may be controlled by the plant hormones gibberellins, and even if not completely controlled, gibberellins certainly have an effect on the development of the leaf hairs. GL1 causes endoreplication, the replication of DNA without subsequent cell division as well as cell expansion. GL1 turns on the expression of a second gene for trichome formation, GL2, which controls the final stages of trichome formation causing the cellular outgrowth.

Arabidopsis thaliana uses the products of inhibitory genes to control the patterning of trichomes, such as TTG and TRY. The products of these genes will diffuse into the lateral cells, preventing them from forming trichomes and in the case of TRY promoting the formation of pavement cells.

Expression of the gene MIXTA, or its analogue in other species, later in the process of cellular differentiation will cause the formation of conical cells over trichomes. MIXTA is a transcription factor.

Stomatal patterning is a much more controlled process, as the stoma effect the plants water retention and respiration capabilities. As a consequence of these important functions, differentiation of cells to form stomata is also subject to environmental conditions to a much greater degree than other epidermal cell types.

Stomata are pores in the plant epidermis that are surrounded by two guard cells, which control the opening and closing of the aperture. These guard cells are in turn surrounded by subsidiary cells which provide a supporting role for the guard cells.

Stomata begin as stomatal meristemoids.[clarification needed] The process varies between dicots and monocots. Spacing is thought to be essentially random in dicots though mutants do show it is under some form of genetic control, but it is more controlled in monocots, where stomata arise from specific asymmetric divisions of protoderm cells. The smaller of the two cells produced becomes the guard mother cells. Adjacent epidermal cells will also divide asymmetrically to form the subsidiary cells.

Because stomata play such an important role in the plants survival, collecting information on their differentiation is difficult by the traditional means of genetic manipulation, as stomatal mutants tend to be unable to survive. Thus the control of the process is not well understood. Some genes have been identified. TMM is thought to control the timing of stomatal initiation specification and FLP is thought to be involved in preventing further division of the guard cells once they are formed.

Environmental conditions affect the development of stomata, in particular their density on the leaf surface. It is thought that plant hormones, such as ethylene and cytokines, control the stomatal developmental response to the environmental conditions. Accumulation of these hormones appears to cause increased stomatal density such as when the plants are kept in closed environments.

Ground Tissue

The ground tissue of plants includes all tissues that are neither dermal nor vascular. It can be divided into three types based on the nature of the cell walls.

- Parenchyma cells have thin primary walls and usually remain alive after they become mature. Parenchyma forms the "filler" tissue in the soft parts of plants, and is usually present in cortex, pericycle, pith, and medullary rays in primary stem and root.

- Collenchyma cells have thin primary walls with some areas of secondary thickening. Collenchyma provides extra mechanical and structural support, particularly in regions of new growth.

- Sclerenchyma cells have thick lignified secondary walls and often die when mature. Sclerenchyma provides the main structural support to a plant.

Parenchyma

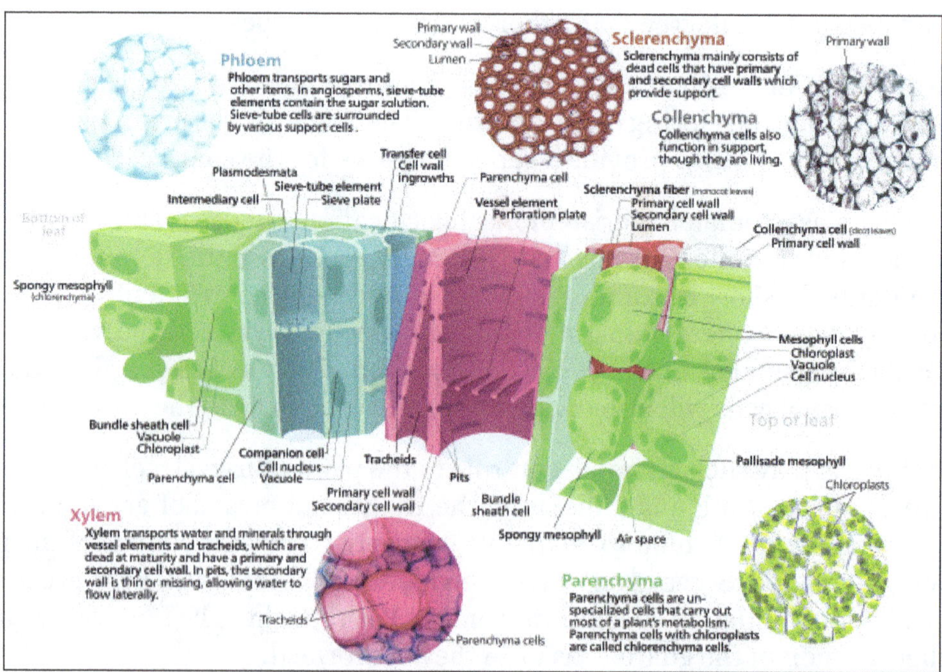

Cross section of a leaf showing various ground tissue types.

Parenchyma is a versatile ground tissue that generally constitutes the "filler" tissue in soft parts of plants. It forms, among other things, the cortex (outer region) and pith (central region) of stems, the cortex of roots, the mesophyll of leaves, the pulp of fruits, and the endosperm of seeds. Parenchyma cells are living cells and may remain meristematic at maturity—meaning that they are capable of cell division if stimulated. They have thin and flexible cellulose cell walls, and are generally polyhedral when close-packed, but can be roughly spherical when isolated from their neighbours. Parenchyma cells are generally large. They have large central vacuoles, which allow the cells to store and regulate ions, waste products, and water. Tissue specialised for food storage is commonly formed of parenchyma cells.

Parenchyma cells have a variety of functions:

- Their main function is to repair.

- In leaves, they form two layers of mesophyll cells immediately beneath the epidermis of the leaf, that are responsible for photosynthesis and the exchange of gases. These layers are called the palisade parenchyma and spongy mesophyll. Palisade parenchyma cells can be either cuboidal or elongated. Parenchyma cells in the mesophyll of leaves are specialised parenchyma cells called chlorenchyma cells (parenchyma cells with chloroplasts). Chlorenchyma cells are also found in other parts of the plant.

- Storage of starch, protein, fats, oils and water in roots, tubers (e.g. potatoes), seed endosperm (e.g. cereals) and cotyledons (e.g. pulses and peanuts).

- Secretion (e.g. the parenchyma cells lining the inside of resin ducts).

- Wound repair and the potential for renewed meristematic activity.

- Other specialised functions such as aeration (aerenchyma) provides buoyancy and helps aquatic plants float.

- Chlorenchyma cells carry out photosynthesis and manufacture food.

The shape of parenchyma cells varies with their function. In the spongy mesophyll of a leaf, parenchyma cells range from near-spherical and loosely arranged with large intercellular spaces, to branched or stellate, mutually interconnected with their neighbours at the ends of their arms to form a three-dimensional network, like in the red kidney bean Phaseolus vulgaris and other mesophytes. These cells, along with the epidermal guard cells of the stoma, form a system of air spaces and chambers that regulate the exchange of gases. In some works, the cells of the leaf epidermis are regarded as specialised parenchymal cells, but the modern preference has long been to classify the epidermis as plant dermal tissue, and parenchyma as ground tissue.

Shapes of parenchyma:

- Polyhedral (found in pallisade tissue of the leaf).

- Circular.

- Stellate (found in stem of plants and have well developed air spaces between them).

- Elongated (also found in pallisade tissue of leaf).

- Lobed (found in spongy and pallisade mesophyll tissue of some plants).

Collenchyma

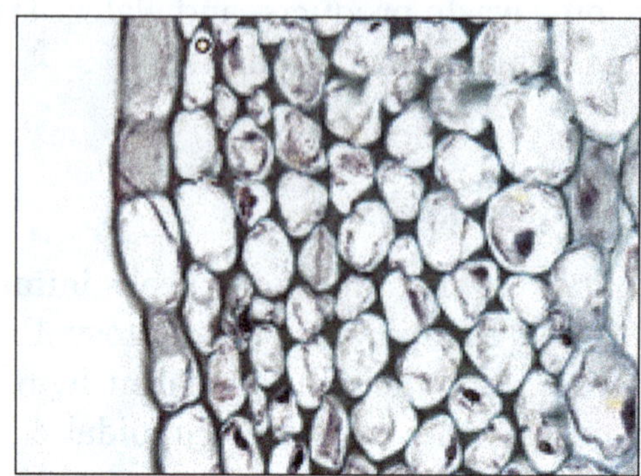
Cross section of collenchyma cells.

Collenchyma tissue is composed of elongated cells with irregularly thickened walls. They provide structural support, particularly in growing shoots and leaves. Collenchyma tissue makes up things such as the resilient strands in stalks of celery. Collenchyma cells are usually living, and have only a thick primary cell wall made up of cellulose and pectin. Cell wall thickness is strongly affected by mechanical stress upon the plant. The walls of collenchyma in shaken plants (to mimic the effects of wind etc)., may be 40–100% thicker than those not shaken.

There are four main types of collenchyma:

- Angular collenchyma (thickened at intercellular contact points).

- Tangential collenchyma (cells arranged into ordered rows and thickened at the tangential face of the cell wall).

- Annular collenchyma (uniformly thickened cell walls).

- Lacunar collenchyma (collenchyma with intercellular spaces).

Collenchyma cells are most often found adjacent to outer growing tissues such as the vascular cambium and are known for increasing structural support and integrity.

Sclerenchyma

Sclerenchyma is the tissue which makes the plant hard and stiff. Sclerenchyma is the supporting tissue in plants. Two types of sclerenchyma cells exist: fibers and sclereids. Their cell walls consist of cellulose, hemicellulose, and lignin. Sclerenchyma cells are the principal supporting cells in plant tissues that have ceased elongation. Sclerenchyma fibers are of great economic importance, since they constitute the source material for many fabrics (e.g. [flax] hemp, jute, and ramie).

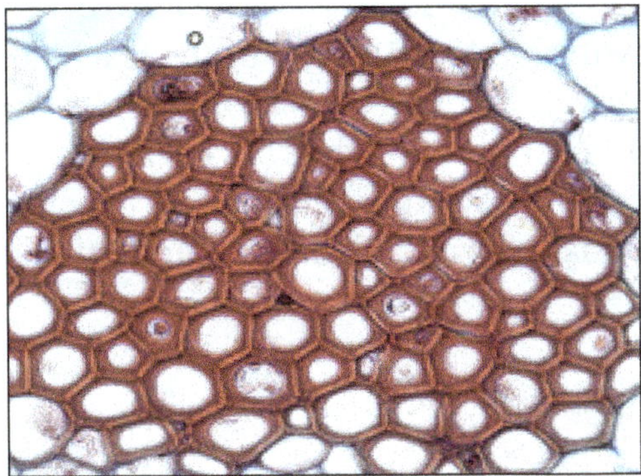

Cross section of sclerenchyma fibers.

Unlike the collenchyma, mature sclerenchyma is composed of dead cells with extremely thick cell walls (secondary walls) that make up to 90% of the whole cell volume. It is the hard, thick walls that make sclerenchyma cells important strengthening and supporting elements in plant parts that have ceased elongation. The difference between fibers and sclereids is not always clear: transitions do exist, sometimes even within the same plant.

Fibers or bast are generally long, slender, so-called prosenchymatous cells, usually occurring in strands or bundles. Such bundles or the totality of a stem's bundles are colloquially called fibers. Their high load-bearing capacity and the ease with which they can be processed has since antiquity made them the source material for a number of things, like ropes, fabrics and mattresses. The fibers of flax (Linum usitatissimum) have been known in Europe and Egypt for more than 3,000 years, those of hemp (Cannabis sativa) in China for just as long. These fibers, and those of jute (Corchorus capsularis) and ramie (Boehmeria nivea, a nettle), are extremely soft and elastic and are especially well suited for the processing to textiles. Their principal cell wall material is cellulose.

Contrasting are hard fibers that are mostly found in monocots. Typical examples are the fiber of many grasses, agaves (sisal: Agave sisalana), lilies (Yucca or Phormium tenax), Musa textilis and others. Their cell walls contain, besides cellulose, a high proportion of lignin. The load-bearing capacity of Phormium tenax is as high as 20–25 kg/ mm², the same as that of good steel wire (25 kg/ mm²), but the fibre tears as soon as too great a strain is placed upon it, while the wire distorts and does not tear before a strain of 80 kg/mm². The thickening of a cell wall has been studied in Linum. Starting at the centre of the fiber, the thickening layers of the secondary wall are deposited one after the other. Growth at both tips of the cell leads to simultaneous elongation. During development the layers of secondary material seem like tubes, of which the outer one is always longer and older than the next. After completion of growth, the missing parts are supplemented, so that the wall is evenly thickened up to the tips of the fibers.

Fibers usually originate from meristematic tissues. Cambium and procambium are their main centers of production. They are usually associated with the xylem and phloem of the vascular bundles. The fibers of the xylem are always lignified, while those of the phloem are cellulosic. Reliable evidence for the fibre cells' evolutionary origin from tracheids exists. During evolution the strength of the tracheid cell walls was enhanced, the ability to conduct water was lost and the size of the pits was reduced. Fibers that do not belong to the xylem are bast (outside the ring of cambium) and such fibers that are arranged in characteristic patterns at different sites of the shoot. The term "sclerenchyma" (originally Sclerenchyma) was introduced by Mettenius in 1865.

Sclereids

Long, tapered sclereids supporting a leaf edge in Dionysia kossinskyi.

Sclereids are the reduced form of sclerenchyma cells with highly thickened, lignified walls. They are small bundles of sclerenchyma tissue in plants that form durable layers, such as the cores of apples and the gritty texture of pears (Pyrus communis). Sclereids are variable in shape. The cells can be isodiametric, prosenchymatic, forked or elaborately branched. They can be grouped into bundles, can form complete tubes located at the periphery or can occur as single cells or small groups of cells within parenchyma tissues. But compared with most fibres, sclereids are relatively short. Characteristic examples are brachysclereids or the stone cells (called stone cells because of their hardness) of pears and quinces (Cydonia oblonga) and those of the shoot of the wax plant (Hoya carnosa). The cell walls fill nearly all the cell's volume. A layering of the walls and the existence of branched pits is clearly visible. Branched pits such as these are called ramiform pits. The shell of many seeds like those of nuts as well as the stones of drupes like cherries and plums are made up from sclereids. These structures are used to protect other cells.

References

- Plant-cell: biologydictionary.net, Retrieved 12 July, 2020

- Nuclear-envelope: lifeofplant.blogspot.com, Retrieved 05 April, 2020

- Cell-wall, plant-anatomy: britannica.com, Retrieved 09 July, 2020

- Sato N (2006). "Origin and Evolution of Plastids: Genomic View on the Unification and Diversity of Plastids". In Wise RR, Hoober JK (eds.). The Structure and Function of Plastids. Advances in Photosynthesis and Respiration. 23. Springer Netherlands. pp. 75–102. doi:10.1007/978-1-4020-4061-0_4. ISBN 978-1-4020-4060-3

- Nucleus, cells/plants: micro.magnet.fsu.edu, Retrieved 17 August, 2020

- Galun, Esra (2007). Plant Patterning: Structural and Molecular Genetic Aspects. World Scientific Publishing Company. p. 333. ISBN 9789812704085

Stem and Root Anatomy

3

- **Stem**
- **Bud**
- **Root**
- **Root**
- **Root Hair**
- **Root Nodule**
- **Root Endodermis**

The stem is the ascending part of the plant which is divided into nodes and internodes. It bears leaves, branches and flowers. Parts of a root include the primary root, lateral roots, the apical meristem, a root cap, and root hair. The topics elaborated in this chapter will help in gaining a better perspective of stem and root anatomy.

Stem

Stem, in botany, the plant axis that bears buds and shoots with leaves and, at its basal end, roots. The stem conducts water, minerals, and food to other parts of the plant; it may also store food, and green stems themselves produce food. In most plants the stem is the major vertical shoot, in some it is inconspicuous, and in others it is modified and resembles other plant parts (e.g., underground stems may look like roots).

Ball cactus (Parodia magnifica).

The primary functions of the stem are to support the leaves; to conduct water and minerals to the leaves, where they can be converted into usable products by photosynthesis; and to transport these products from the leaves to other parts of the plant, including the roots. The stem conducts water and nutrient minerals from their site of absorption in the roots to the leaves by means of certain vascular tissues in the xylem. The movement of synthesized foods from the leaves to other plant organs occurs chiefly through other vascular tissues in the stem called phloem. Food and water are also frequently stored in the stem. Examples of food-storing stems include such specialized forms as tubers, rhizomes, and corms and the woody stems of trees and shrubs. Water storage is developed to a high degree in the stems of cacti, and all green stems are capable of photosynthesis.

Leaves, stem, and root system of a fig tree seedling (Ficus).

Growth and Anatomy

The first rudiment of the young stem, or shoot, of an embryonic plant appears from the seed after the root has first protruded. The growing portion at the apex of the shoot is the terminal bud of the plant, and by the continued development of this bud and its adjacent tissues, the stem increases in height. Lateral buds and leaves grow out of the stem at intervals called nodes; the intervals on the stem between the nodes are called internodes. The number of leaves that appear at a node depends on the species of plant; one leaf per node is common, but two or or more leaves may grow at the nodes of some species. When a leaf drops off a stem at the end of a growing season, it leaves a scar on the stem because of the severing of the vascular (conducting) bundles that had connected stem and leaf. As the stem continues to grow, lateral buds are produced that develop into lateral shoots more or less resembling the parent stem, and these ultimately determine the branching of the plant. In trees the lateral shoots develop into branches, from which other lateral shoots, called branchlets, or twigs, arise. The point at which a leaf diverges in axis from a stem is called the axil. A bud formed in the axil of a previously formed leaf is called an axillary bud, and it, like the leaves, is produced from the tissues of the stem. During the development of such buds, vascular bundles are formed within them that are continuous with those of the stem.

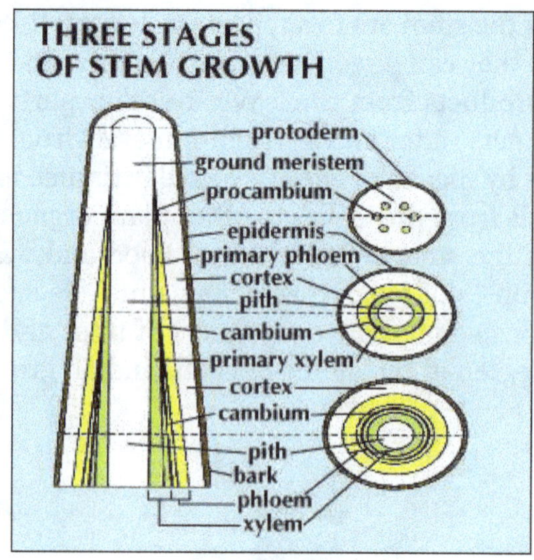

Stem anatomy.

In the stems of young dicotyledons (angiosperms with two seed leaves) and gymnosperms, the vascular bundles (xylem and phloem) are arranged in a circle around a central core of spongy ground tissue called the pith. Surrounding the vascular bundles is a layer that varies in thickness in different species and is called the cortex. Surrounding this and comprising the exterior surface of the stem is a layer called the epidermis. In plants with woody stems, a variety of secondary tissues are added to these primary tissues. Among the most important of these is a ring of meristematic cells that in turn give rise to the vascular cambium. This tissue arises between the primary xylem and phloem and gives rise to secondary phloem on the outside and secondary xylem on the inside; the latter tissue is the wood of trees.

Stem Types and Modifications

Joshua tree.

Many plants are annuals and complete their life cycles in one growing season, after which the entire plant, including the stem, dies. In biennial plants the lower part of the stem, often modified for food storage, persists after the first growing season and bears buds from which an erect stem arises during the second growing season. In perennial plants the short stem may produce new shoots for many years. Plants producing woody stems are called trees and shrubs; the latter produce branches from or near the ground, while the former have conspicuous trunks.

In general, the habit of a stem is erect or ascending, but it may lie prostrate on the ground, as in the sweet potato and strawberry. A stem may climb on rocks or plants by means of rootlets, as in ivy; other vines have twining stems that twist around a supporting plant in a spiral manner, as in the honeysuckle and hop. In other cases, climbing plants are supported by tendrils that may be specialized stems, as in the grape and passion-flower. In tropical climates twining plants often form thick woody stems and are called lianas, while in temperate regions they are generally herbaceous vines. A stolon is a stem that curves toward the ground and, on reaching a moist spot, takes root and forms an upright stem and ultimately a separate plant. Among the subterranean stems are the rhizome, corm, and tuber. In some plants the stem does not elongate during its early development but instead forms a short conical structure from which a crown of leaves arises. These may form a bulb (as in the onion and lily), a head (cabbage, lettuce), or a rosette (dandelion, plantain).

Twining stem of a decorative vine.

Tuber.

Bud

In botany, a bud is an undeveloped or embryonic shoot and normally occurs in the axil of a leaf or at the tip of a stem. Once formed, a bud may remain for some time in a dormant condition, or it may form a shoot immediately. Buds may be specialized to develop flowers or short shoots, or may have the potential for general shoot development.

The term bud is also used in zoology, where it refers to an outgrowth from the body which can develop into a new individual.

Inflorescence bud scales in Halesia carolina.

The buds of many woody plants, especially in temperate or cold climates, are protected by a covering of modified leaves called scales which tightly enclose the more delicate parts of the bud. Many bud scales are covered by a gummy substance which serves as added protection. When the bud develops, the scales may enlarge somewhat but usually just drop off, leaving a series of horizontally-elongated scars on the surface of the growing stem. By means of these scars one can determine the age of any young branch, since each year's growth ends in the formation of a bud, the formation of which produces an additional group of bud scale scars. Continued growth of the branch causes these scars to be obliterated after a few years so that the total age of older branches cannot be determined by this means.

European beech (Fagus sylvatica) bud.

In many plants scales do not form over the bud, and the bud is then called a naked bud. The minute underdeveloped leaves in such buds are often excessively hairy. Naked buds are found in some shrubs, like some species of the Sumac and Viburnums (Viburnum alnifolium and V. lantana) and in herbaceous plants. In many of the latter, buds are even more reduced, often consisting of undifferentiated masses of cells in the axils of leaves. A terminal bud occurs on the end of a stem and lateral buds are found on the side. A head of cabbage is an exceptionally large terminal bud, while Brussels sprouts are large lateral buds.

Since buds are formed in the axils of leaves, their distribution on the stem is the same as that of leaves. There are alternate, opposite, and whorled buds, as well as the terminal bud at the tip of the stem. In many plants buds appear in unexpected places: these are known as adventitious buds.

Often it is possible to find a bud in a remarkable series of gradations of bud scales. In the buckeye, for example, one may see a complete gradation from the small brown outer scale through larger scales which on unfolding become somewhat green to the inner scales of the bud, which are remarkably leaf-like. Such a series suggests that the scales of the bud are in truth leaves, modified to protect the more delicate parts of the plant during unfavorable periods.

Types of Buds

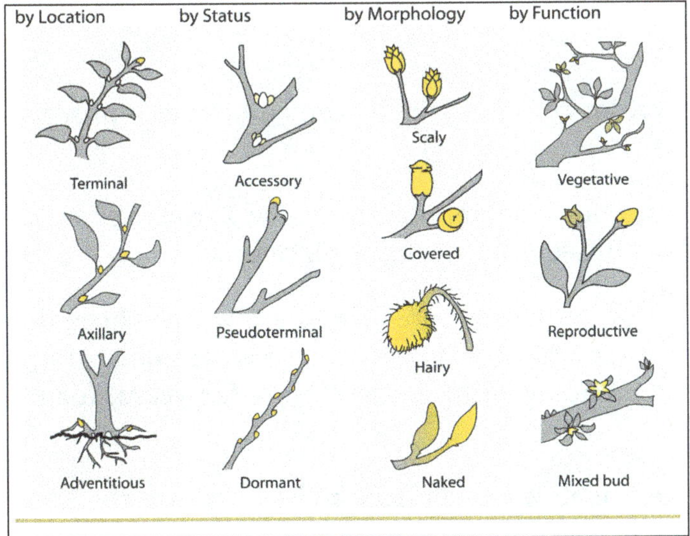

Plant buds classification.

Terminal, vegetative bud of Ficus carica.

Buds are often useful in the identification of plants, especially for woody plants in winter when leaves have fallen. Buds may be classified and described according to different criteria: location, status, morphology, and function.

Botanists commonly use the following terms:

- For location:

  ○ Terminal, when located at the tip of a stem (apical is equivalent but rather reserved for the one at the top of the plant);

  ○ Axillary, when located in the axil of a leaf (lateral is the equivalent but some adventitious buds may be lateral too);

  ○ Adventitious, when occurring elsewhere, for example on trunk or on roots (some adventitious buds may be former axillary ones reduced and hidden under the bark, other adventitious buds are completely new formed ones).

- For status:

  ○ Accessory, for secondary buds formed besides a principal bud (axillary or terminal);

  ○ Resting, for buds that form at the end of a growth season, which will lie dormant until onset of the next growth season;

  ○ Dormant or latent, for buds whose growth has been delayed for a rather long time. The term is usable as a synonym of resting, but is better employed for buds waiting undeveloped for years, for example epicormic buds;

  ○ Pseudoterminal, for an axillary bud taking over the function of a terminal bud (characteristic of species whose growth is sympodial: terminal bud dies and is replaced by the closer axillary bud, for examples beech, persimmon, Platanus have sympodial growth).

- For morphology:

  ○ Scaly or covered (perulate), when scales, also referred to as a perule (lat. perula, perulaei) (which are in fact transformed and reduced leaves) cover and protect the embryonic parts;

  ○ Naked, when not covered by scales;

  ○ Hairy, when also protected by hairs (it may apply either to scaly or to naked buds).

- For function:

  ○ Vegetative, if only containing vegetative pieces: embryonic shoot with leaves (a leaf bud is the same);

  ○ Reproductive, if containing embryonic flowers (a flower bud is the same);

  ○ Mixed, if containing both embryonic leaves and flowers.

Root

Root, in botany, that part of a vascular plant normally underground. Its primary functions are anchorage of the plant, absorption of water and dissolved minerals and conduction of these to the stem, and storage of reserve foods. The root differs from the stem mainly by lacking leaf scars and buds, having a root cap, and having branches that originate from internal tissue rather than from buds.

Watercress seedling.

Types of Roots and Root Systems

The primary root, or radicle, is the first organ to appear when a seed germinates. It grows downward into the soil, anchoring the seedling. In gymnosperms and dicotyledons (angiosperms with two seed leaves), the radicle becomes a taproot. It grows downward, and secondary roots grow laterally from it to form a taproot system. In some plants, such as carrots and turnips, the taproot also serves as food storage.

Two types of root system: (left) the fibrous roots of grass and
(right) the fleshy taproot of a sugar beet.

Grasses and other monocotyledons (angiosperms with a single seed leaf) have a fibrous root system, characterized by a mass of roots of about equal diameter. This network of roots does not arise as branches of the primary root but consists of many branching roots that emerge from the base of the stem.

Banyan tree (Ficus species) with aerial roots emerging from the branches.

Some roots, called adventitious roots, arise from an organ other than the root—usually a stem, sometimes a leaf. They are especially numerous on underground stems, such as rhizomes, corms, and tubers, and make it possible to vegetatively propagate many plants from stem or leaf cuttings. Certain adventitious roots, known as aerial roots, either pass for some distance through the air before reaching the soil or remain hanging in the air. Some of these, such as those seen in corn (maize), screw pine, and banyan, eventually assist in supporting the plant in the soil. In many epiphytic plants, such as various orchids and Tillandsia species, aerial roots are the primary means of attachment to non-soil surfaces such as other plants and rocks.

A number of other specialized roots exist among vascular plants. Pneumatophores, commonly found in mangrove species that grow in saline mud flats, are lateral roots that grow upward out of the mud and water to function as the site of oxygen intake for the submerged primary root system. The roots of certain parasitic plants are highly modified into haustoria, which embed into the vascular system of the host plant to feed the parasite. The nodular roots of many members of the pea family (Fabaceae) host symbiotic nitrogen-fixing bacteria, and many plant roots also form intricate associations with mycorrhizal soil fungi; a number of non-photosynthetic mycoheterotrophic plants, such as Indian pipe, rely exclusively on these fungi for nutrition.

Pneumatophores of the black mangrove (Avicennia germinans) encrusted with salt and a young seedling projecting above the surface of the water.

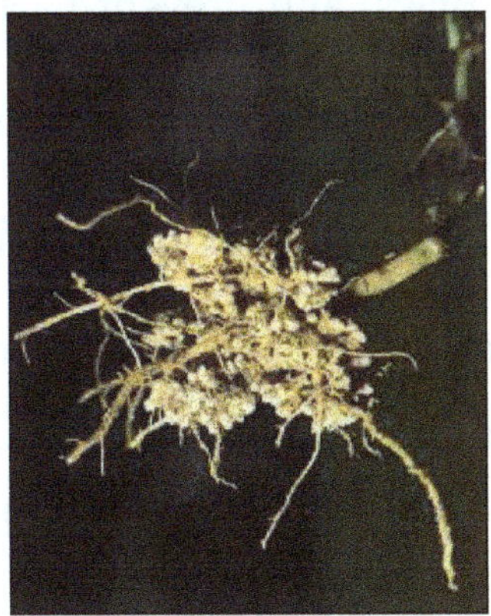

The roots of an Austrian winter pea plant (Pisum sativum) with nodules harbouring nitrogen-fixing bacteria (Rhizobium). Root nodules develop as a result of a symbiotic relationship between rhizobial bacteria and the root hairs of the plant.

Root Hair

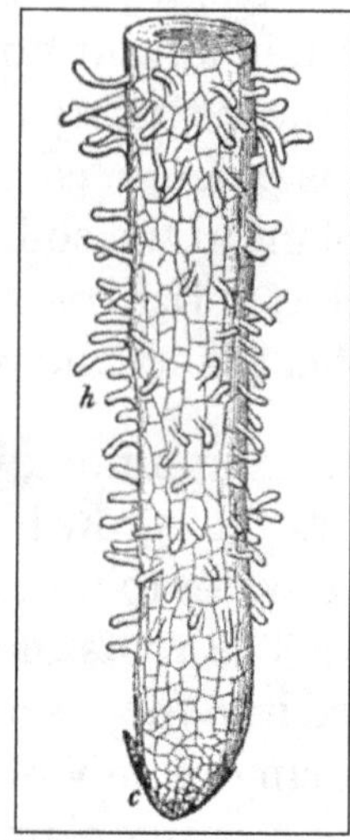

Drawing of root tip, showing young root hairs.

A root hair, or absorbent hair, the rhizoid of a vascular plant, is a tubular outgrowth of a trichoblast, a hair-forming cell on the epidermis of a plant root. As they are lateral extensions of a single cell and only rarely branched, they are visible to the naked eye and light microscope. They are found only in the region of maturation of the root. Just prior to, and during, root hair cell development, there is elevated phosphorylase activity. Plants absorb water from the soil by osmosis. Root hair cells are adapted for this by having a large surface area to speed up osmosis. Another adaptation that they have is root hair cells have a large permanent vacuole.

Function

The root hairs are where most water absorption happens. They are long and so they can penetrate between soil particles, and prevent harmful bacterial organisms from entering the plant through the xylem vessels. They have a large surface area for absorption of water. Increasing root surface area aids plants to be more efficient in absorbing nutrients and establishing relationships with microbes. Cross-section of root hair cell: a roughly rectangular shape with a long, thin tail extending to the right and a nucleus at the top left. Water passes from the soil water to the root hair cell's cytoplasm by osmosis. This happens because the soil water has a higher water potential than the root hair cell cytoplasm. The function of root hairs is to collect water and mineral nutrients that are present in the soil and take this solution up through the roots to the rest of the plant. As root hair cells do not carry out photosynthesis they do not contain chloroplasts.

Formation

Root hair cells are outgrowths at a tip of the plant's roots. Root hair cells vary between

15 and 17 micrometres in diameter, and 80 to 1,500 micrometres in length. They are found only in the zone of maturation, and not the zone of elongation, possibly because any root hairs that arise are sheared off as the root elongates and moves through the soil. Root hairs grow quickly, at least 1μm/min, making them particularly useful for research on cell expansion.

Importance

Root hairs form an important surface as they are needed to absorb most of the water and nutrients needed for the plant. They are also directly involved in the formation of root nodules in legume plants. The root hairs curl around the bacteria which allows for the formation of an infection thread through into the dividing cortical cells to form the nodule.

Having a large surface area, the active uptake of water and minerals through root hairs is highly efficient. Root hair cells also secrete acid (H+ from malic acid) which exchanges and helps solubilize the minerals into ionic form, making the ions easier to absorb.

Survival

Root hair cells can survive for 2 to 3 weeks and then die off; at the same time, new root hair cells are continually being formed at the tip of the root. This way, the root hair coverage stays the same. When a new root hair cell grows, it excretes a poison so that the other cells in close proximity to it are unable to grow one of these hairs. This ensures equal and efficient distribution of the actual hairs on these cells.

The act of re-potting or transplanting a plant can result in root hair cells being pulled off, perhaps to a significant extent, and such plants may therefore wilt for some time as a result.

Root Nodule

Root nodules are found on the roots of plants, primarily legumes, that form a symbiosis with nitrogen-fixing bacteria. Under nitrogen-limiting conditions, capable plants form a symbiotic relationship with a host-specific strain of bacteria known as rhizobia. This process has evolved multiple times within the legumes, as well as in other species found within the Rosid clade. Legume crops include beans, peas, and soybeans.

Within legume root nodules, nitrogen gas (N_2) from the atmosphere is converted into ammonia (NH_3), which is then assimilated into amino acids (the building blocks of proteins), nucleotides (the building blocks of DNA and RNA as well as the important energy molecule ATP), and other cellular constituents such as vitamins, flavones, and hormones. Their ability to fix gaseous nitrogen makes legumes an ideal agricultural organism as their requirement for nitrogen fertilizer is reduced. Indeed, high nitrogen content blocks nodule development as there is no benefit for the plant of forming the symbiosis. The energy for splitting the nitrogen gas in the nodule comes from sugar that is translocated from the leaf (a product of photosynthesis). Malate as a breakdown product of sucrose is the direct carbon source for the bacteroid. Nitrogen fixation in the nodule is very oxygen sensitive. Legume nodules harbor an iron containing protein called leghaemoglobin, closely related to animal myoglobin, to facilitate the diffusion of oxygen gas used in respiration.

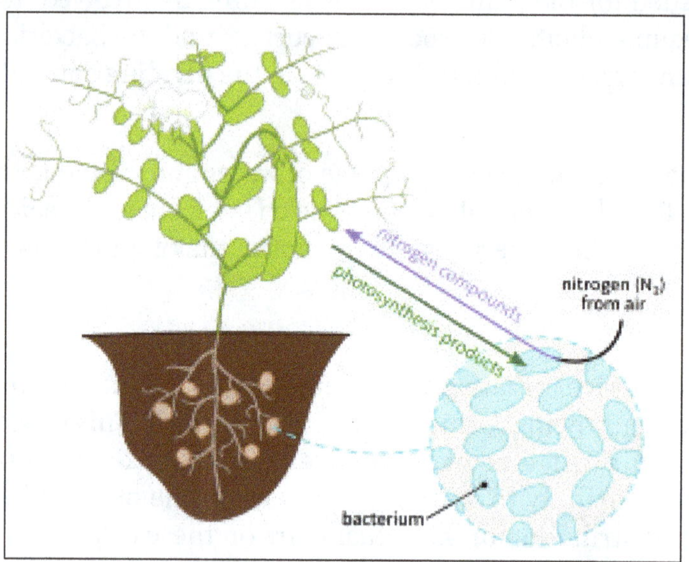

A simplified diagram of the relation between the plant and the symbiotic bateria (cyan) in the root nodules.

Symbiosis

Leguminous Family

Plants that contribute to nitrogen fixation include the legume family – Fabaceae – with taxa such as kudzu, clovers, soybeans, alfalfa, lupines, peanuts, and rooibos. They contain symbiotic bacteria called rhizobia within the nodules, producing nitrogen compounds that help the plant to grow and compete with other plants. When the plant dies, the fixed nitrogen is released, making it available to other plants and this helps to fertilize the soil. The great majority of legumes have this association, but a few genera (e.g., Styphnolobium) do not. In many traditional and organic farming practices, fields are rotated through various types of crops, which usually includes one consisting mainly or entirely of clover or buckwheat (non-legume family Polygonaceae), which are often referred to as "green manure".

Inga alley farming relies on the leguminous genus Inga, a small tropical, tough-leaved, nitrogen-fixing tree.

Non-leguminous

Although by far the majority of plants able to form nitrogen-fixing root nodules are in the legume family Fabaceae, there are a few exceptions:

- Parasponia, a tropical genus in the Cannabaceae also able to interact with rhizobia and form nitrogen-fixing nodules.

- Actinorhizal plants such as alder and bayberry can also form nitrogen-fixing nodules, thanks to a symbiotic association with Frankia bacteria. These plants belong to 25 genera distributed among 8 plant families.

The ability to fix nitrogen is far from universally present in these families. For instance, of 122 genera in the Rosaceae, only 4 genera are capable of fixing nitrogen. All these families belong to the orders Cucurbitales, Fagales, and Rosales, which together with the Fabales form a clade of eurosids. In this clade, Fabales were the first lineage to branch off; thus, the ability to fix nitrogen may be plesiomorphic and subsequently lost in most descendants of the original nitrogen-fixing plant; however, it may be that the basic genetic and physiological requirements were present in an incipient state in the last common ancestors of all these plants.

Classification

Two main types of nodule have been described: determinate and indeterminate.

Determinate nodules are found on certain tribes of tropical legume such as those of the genera Glycine (soybean), Phaseolus (common bean), and Vigna. and on some temperate legumes such as Lotus. These determinate nodules lose meristematic activity shortly after initiation, thus growth is due to cell expansion resulting in mature nodules which are spherical in shape. Another types of determinate nodule is found in a wide range of herbs, shrubs and trees, such as Arachis (peanut). These are always associated with the axils of lateral or adventitious roots and are formed following infection via cracks where these roots emerge and not using root hairs. Their internal structure is quite different from those of the soybean type of nodule.

Indeterminate nodules are found in the majority of legumes from all three sub-families, whether in temperate regions or in the tropics. They can be seen in Faboideae legumes such as Pisum (pea), Medicago (alfalfa), Trifolium (clover), and Vicia (vetch) and all mimosoid legumes such as acacias, the few nodulated caesalpinioid legumes such as partridge pea. They earned the name "indeterminate" because they maintain an active apical meristem that produces new cells for growth over the life of the nodule. This results in the nodule having a generally cylindrical shape, which may be extensively branched. Because they are actively growing, indeterminate nodules manifest zones which demarcate different stages of development/symbiosis:

- Zone I—the active meristem. This is where new nodule tissue is formed which will later differentiate into the other zones of the nodule.

- Zone II—the infection zone. This zone is permeated with infection threads full of bacteria. The plant cells are larger than in the previous zone and cell division is halted.

- Interzone II–III—Here the bacteria have entered the plant cells, which contain amyloplasts. They elongate and begin terminally differentiating into symbiotic, nitrogen-fixing bacteroids.

- Zone III—the nitrogen fixation zone. Each cell in this zone contains a large, central vacuole and the cytoplasm is filled with fully differentiated bacteroids which are actively fixing nitrogen. The plant provides these cells with leghemoglobin, resulting in a distinct pink color.

- Zone IV—the senescent zone. Here plant cells and their bacteroid contents are being degraded. The breakdown of the heme component of leghemoglobin results in a visible greening at the base of the nodule. This is the most widely studied type of nodule, but the details are quite different in nodules of peanut and relatives and some other important crops such as lupins where the nodule is formed following direct infection of rhizobia through the epidermis and where infection threads are never formed. Nodules grow around the root, forming a collar-like structure. In these nodules and in the peanut type the central infected tissue is uniform, lacking the uninfected ells seen in nodules of soybean and many indeterminate types such as peas and clovers.

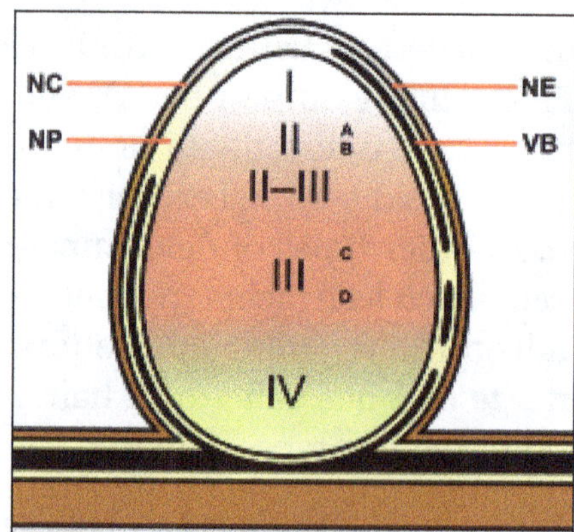

Diagram illustrating the different zones of an indeterminate root nodule.

Nodulation

Legumes release organic compounds as secondary metabolites called flavonoids from their roots, which attract the rhizobia to them and which also activate nod genes in the bacteria to produce nod factors and intitiate nodule formation. These nod factors initiate root hair curling. The curling begins with the very tip of the root hair curling around the Rhizobium. Within the root tip, a small tube called the infection thread forms, which provides a pathway for the Rhizobium to travel into the root epidermal cells as the root hair continues to curl.

Partial curling can even be achieved by nod factor alone. This was demonstrated by the isolation of nod factors and their application to parts of the root hair. The root hairs curled in the direction of the application, demonstrating the action of a root hair attempting to curl around a bacterium. Even application on lateral roots caused curling. This demonstrated that it is the nod factor itself, not the bacterium that causes the stimulation of the curling.

When the nod factor is sensed by the root, a number of biochemical and morphological changes happen: cell division is triggered in the root to create the nodule, and the root hair growth is redirected to curl around the bacteria multiple times until it fully encapsulates one or more bacteria. The bacteria encapsulated divide multiple times, forming a microcolony. From this microcolony, the bacteria enter the developing nodule through the infection thread, which grows through the root hair into the basal part of the epidermis cell, and onwards into the root cortex; they are then surrounded by a plant-derived symbiosome membrane and differentiate into bacteroids that fix nitrogen.

Effective nodulation takes place approximately four weeks after crop planting, with the size, and shape of the nodules dependent on the crop. Crops such as soybeans, or peanuts will have larger nodules than forage legumes such as red clover, or alfalfa, since their nitrogen needs are higher. The number of nodules, and their internal color, will indicate the status of nitrogen fixation in the plant.

Nodulation is controlled by a variety of processes, both external (heat, acidic soils, drought, nitrate) and internal (autoregulation of nodulation, ethylene). Autoregulation of nodulation controls nodule numbers per plant through a systemic process involving the leaf. Leaf tissue senses the early nodulation events in the root through an unknown chemical signal, then restricts further nodule development in newly developing root tissue. The Leucine rich repeat (LRR) receptor kinases (NARK in soybean (Glycine max); HAR1 in Lotus japonicus, SUNN in Medicago truncatula) are essential for autoregulation of nodulation (AON). Mutation leading to loss of function in these AON receptor kinases leads to supernodulation or hypernodulation. Often root growth abnormalities accompany the loss of AON receptor kinase activity, suggesting that nodule growth and root development are functionally linked. Investigations into the mechanisms of nodule formation showed that the ENOD40 gene, coding for a 12–13 amino acid protein, is up-regulated during nodule formation.

Connection to Root Structure

Root nodules apparently have evolved three times within the Fabaceae but are rare outside that family. The propensity of these plants to develop root nodules seems to relate to their root structure. In particular, a tendency to develop lateral roots in response to abscisic acid may enable the later evolution of root nodules.

Root Endodermis

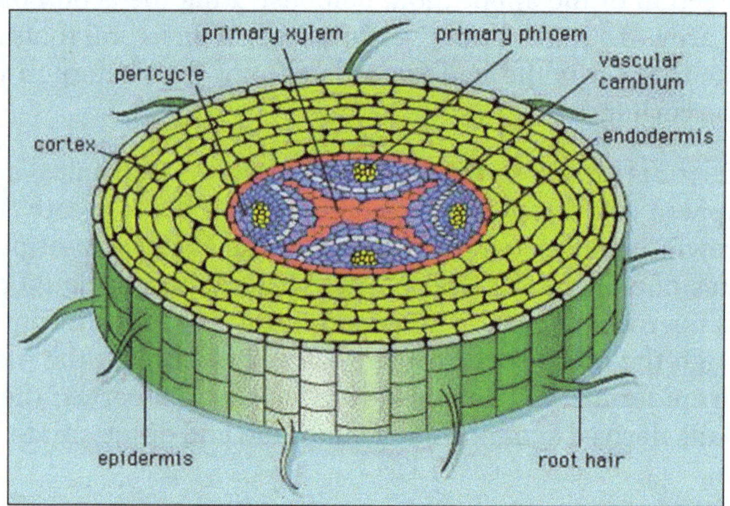

The root endodermis is the cylindrical boundary that separates the inner vascular tissue from the outer cortex and functions as an apoplasmic barrier for selective nutrient uptake. Recent developmental and cell biological studies have started to reveal the mechanisms by which this single cell layer serves as a key regulatory module of root growth, tissue patterning and nutrient flow, which in concert support the plant's ability to survive in a terrestrial habitat.

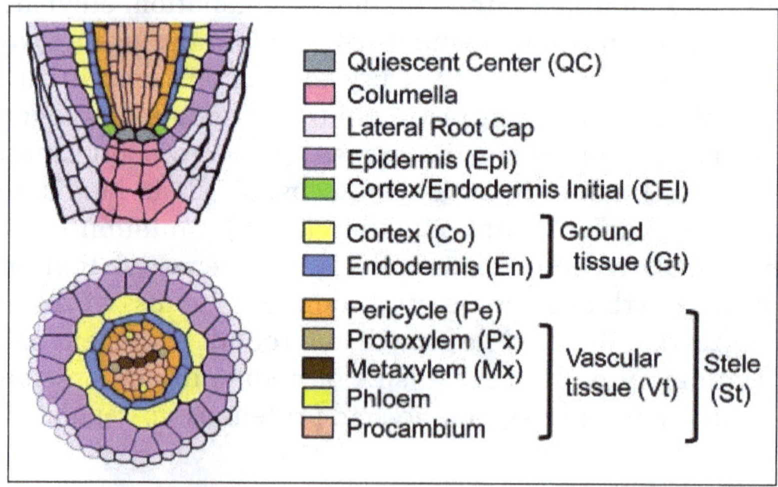

Schematic representation of Arabidopsis root tissue organization.

The roots of vascular plants are composed of functionally distinct tissues organized into a concentric pattern of cell layers. Although both the number and morphology of each tissue vary with plant species, the overall organization of root tissue patterns are well conserved, indicating their importance for basic root functions, such as anchorage of the plant and nutritional uptake. Figure shows a schematic view of the Arabidopsis root tissue organization. The roots of certain plant species, such as rice, maize and tobacco, exhibit more complex patterns, mainly due to the presence of additional cortical layers.

Laser ablation and mosaic analyses indicated that stem cell maintenance and tissue patterning are regulated by cell-cell communication. Recent molecular genetic studies using the Arabidopsis root as a model system have started to reveal the molecular identity of the intercellular signals involved in root growth and tissue patterning, and highlight the importance of the endodermis as a central regulator of these processes. Moreover, a search for genes specifically expressed in the endodermis has led to the identification of the elusive components of the Casparian strip (CS), the cell wall structure that serves as an impermeable apoplasmic barrier to nutrients and water. This minireview introduces recent advances in our understanding of the mechanisms by which the root endodermis differentiates and how this single-celled layer controls root growth and tissue patterning.

Formation of the Root Endodermis

In Arabidopsis, the developmental origin of the root endodermis can be traced back to triangular-stage embryos, where the ground meristem, the progenitor of the ground tissue (endodermis and cortex), divides periclinally to form two-layered ground tissue primordia. After germination, this two-layered ground tissue organization is maintained by the stereotypical cell division of the cortex/endodermis initials (CEIs) and their immediate daughter cells (CEIDs). An unidentified signal that possibly emanates from the adjacent quiescent center (QC) maintains the pluripotent stem cell capacity of the CEI. The CEI divides transversely to give rise to a CEID at the distal position from the QC, whereas the daughter cell abutting the QC maintains its CEI identity. The CEID divides periclinally to give rise to two daughter cells, each of which differentiates into either an endodermis or cortex cell, depending on its position relative to the stele.

Two GRAS-type transcription factors, SHORTROOT (SHR) and SCARECROW (SCR), play key roles in the formation and maintenance of the root endodermis. SHR proteins are produced in the stele and move to the adjacent cell layer, which is composed of the QC, CEI, CEID and endodermis. In this layer, SHR activates the transcription of several target genes, including SCR, and the SCR protein thus formed physically interacts with SHR, likely forming a transcription activation complex. In the CEID, the SHR-SCR complex activates the transcription of a cell cycle regulator, Cyclin D6, which in turn is required for the periclinal division of CEID. Thus, loss-of-function shr and scr mutants are unable to form the two-layered ground tissue. The ground tissue of the shr mutant does not possess the differentiated attributes of the endodermis, whereas that of scr contains at least partial characteristics of the endodermis. Therefore, the functioning of SHR, but not of SCR, is required for the differentiation of the endodermis. Consistent with this observation, the forced expression of SHR outside the stele results in the ectopic manifestation of endodermal characteristics.

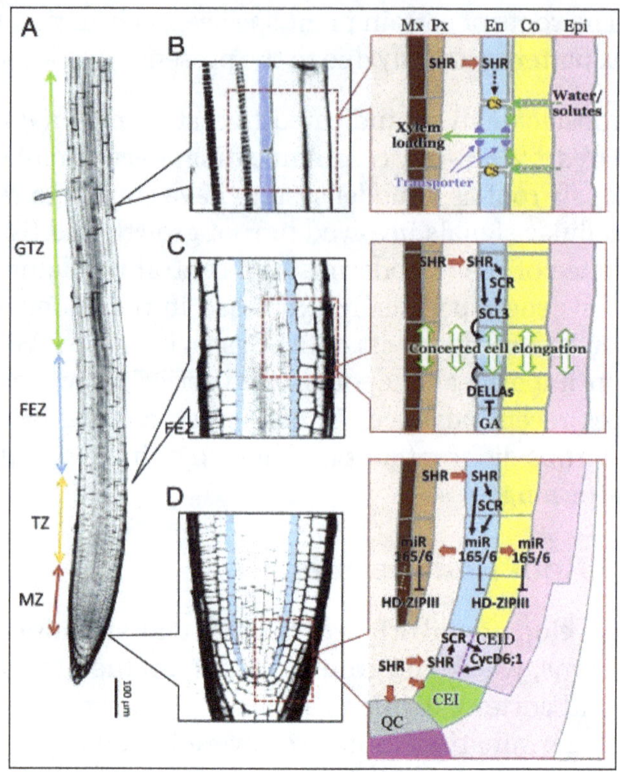

Developmental zones of the Arabidopsis root and the signal transduction pathways and nutrient flow in the region of the root endodermis. (A) A confocal section of an Arabidopsis root. The four developmental zones, i.e., the meristematic zone (MZ), transition zone (TZ), fast elongation zone (FEZ), and growth terminating zone (GTZ) are labeled. (B-D) Magnified views of selected root regions (left; blue color indicates endodermis) and schematic representations of water/solute flow and developmental signaling in the endodermis (right). (B) Formation of the CS (represented as orange dots) and its role in regulating water and solute uptake. (C) GA- and SCL3-mediated control of endodermal cell elongation and its effect on the elongation of adjacent cell layers. (D) SHR- and miR165/6-mediated intercellular signaling between the endodermis and surrounding tissues. Arrows indicate activation/promotion, and T bars represent inhibition/suppression. Thick red arrows indicate cell-cell trafficking.

Whereas the number of cortical layers varies with plant species and the age of the root, the endodermis is typically composed of a single cell layer. This evolutionarily and developmentally conserved property of the root endodermis is likely due to the restricted cell-cell movement of SHR to one cell distance. Loss of SCR protein results in increased SHR movement and ectopic periclinal divisions, indicating that SCR captures SHR protein in the nucleus of the recipient cell layer, and thereby inhibits further cell-cell movement of SHR. Interestingly, this regulatory mechanism seems to be conserved in rice, indicating that the mechanism that ensures the formation of the single endodermis layer is evolutionarily conserved.

References

- Stem-plant, science: britannica.com, Retrieved 13 July, 2020

- Grierson, Claire; Schiefelbein, John (2002-04-04). "Root Hairs". The Arabidopsis Book / American Society of Plant Biologists. 1: e0060. doi:10.1199/tab.0060. ISSN 1543-8120. PMC 3243358. PMID 22303213

- Fabrice Foucher & Eva Kondorosi (2000). "Cell cycle regulation in the course of nodule organogenesis in Medicago". Plant Molecular Biology. 43 (5–6): 773–786. doi:10.1023/A:1006405029600. PMID 11089876

- Miyashima S, Koi S, Hashimoto T, Nakajima K. Non-cell-autonomous microRNA165 acts in a dose-dependent manner to regulate multiple differentiation status in the Arabidopsis root. Development. 2011;138:2303–13. doHelariutta Y, Fukaki H, Wysocka-Diller J, Nakajima K, Jung J, Sena G, et al. The SHORT-ROOT gene controls radial patterning of the Arabidopsis root through radial signaling. Cell. 2000;101:555–67. doi: 10.1016/S0092-8674(00)80865-X

Flower and Leaf Anatomy

4

- **Flower Anatomy**
- **Stamen**
- **Androecium**
- **Gynoecium**
- **Stigma**
- **Sepal**
- **Petal**
- **Perianth**

Flower is the reproductive part in flowering plants. It mainly consists of four parts which are sepals, petals, stamens and carpels. The leaf is the photosynthetic organ of the plant. It comprises blade, petiole and leaflets. The chapter discusses the anatomy of flower and leaf in detail.

Flower Anatomy

It is a plant organ. It is a reproductive structure of an angiospermous plant. It is a determinate, modified shoot having either or both a stamen and pistil where male and female sexual gametes are produced, respectively. It is the pistil which develops into a seed or seeds enclosed in fruits. In addition, this plant organ may also consist of outer whorls of petals and sepals.

It is essential to natural reproduction in the angiosperms by seeds. However, this function is indirect. It does not give rise to a new plant, but develops into a fruit and seed. It is the seed, specifically the embryo inside, which germinates and transforms into a new plant.

Should any angiospermous plant be prevented from producing this particular organ, there would be no fruits and seeds. Stated another way, it is necessary that in crop agriculture where the economic product is the fruit and the seed, angiospermous plants should be provided with the right conditions that will induce floral formation and ensure development.

Flowers are present in various forms and sizes in the angiosperms although they may be small and inconspicuous in some species. This reproductive organ is unique and is the reason why the angiosperms are also called true flowering plants. In contrast, other plants such as the pines, conifers, gnetum, ferns, fern allies, and bryophytes do not have flowers and are excluded from the angiosperms.

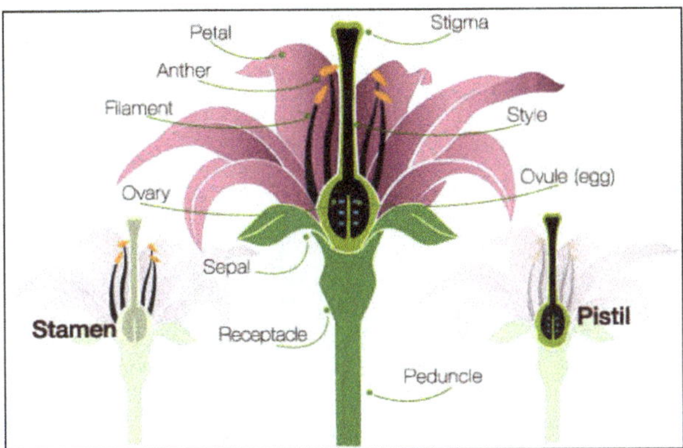

Stamen

Stamen is the male reproductive part of a flower. In all but a few extant angiosperms, the stamen consists of a long slender stalk, the filament, with a two-lobed anther at the tip. The anther consists of four saclike structures (microsporangia) that produce pollen for pollination. Small secretory structures, called nectaries, are often found at the base of the stamens; they provide food rewards for insect and bird pollinators. All the stamens of a flower are collectively called the androecium.

Close-up of a spring crocus (Crocus vernus). The flower's three pollen-covered stamens are clearly visible.

Reproduction in flowering plants begins with pollination, the transfer of pollen from anther to stigma on the same flower or to the stigma of another flower on the same plant (self-pollination) or from the anther on one plant to the stigma of another plant (cross-pollination). Once the pollen grain lodges on the stigma, a pollen tube grows from the pollen grain to an ovule. Two sperm nuclei then pass through the pollen tube. One of them unites with the egg nucleus and produces a zygote. The other sperm nucleus unites with two polar nuclei to produce an endosperm nucleus. The fertilized ovule develops into a seed.

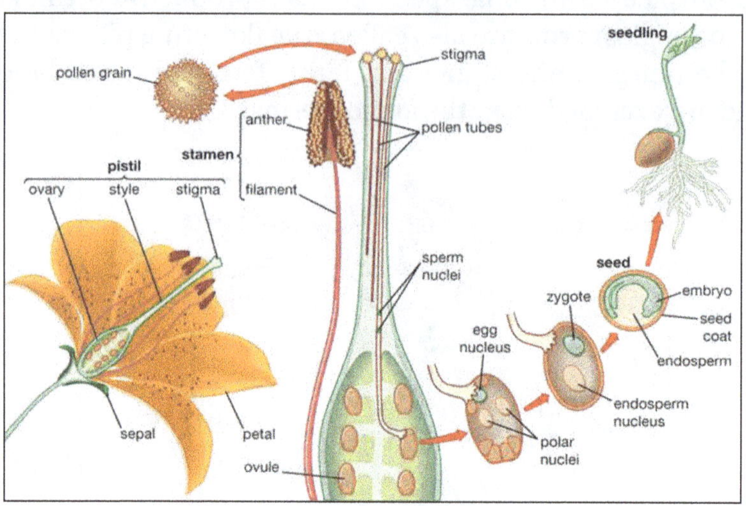

How flowering plants reproduce.

The number and arrangement of stamens, as well as the way in which the anthers release pollen, are important taxonomic characteristics for many flowering plants. The number of stamens is often the same as the number of petals. The presence of numerous stamens is common in many plant families (e.g., Cactaceae, Ranunculaceae, and Rosaceae); most orchids possess only one stamen. In plants with imperfect (unisexual) flowers, the staminate flowers may be borne individually, as in most squash species, or arranged in long clusters known as catkins, as is characteristic of oaks and willows. While the anthers of most angiosperms release pollen through a rupture along one side of each sac, the anthers belonging to members of the heath family (Ericaceae) release pollen through small pores at the anther tip. Some flowers produce sterile stamens, known as staminodes, which may be showy (e.g., on the cannonball tree) or inconspicuous (e.g., in Penstemon species).

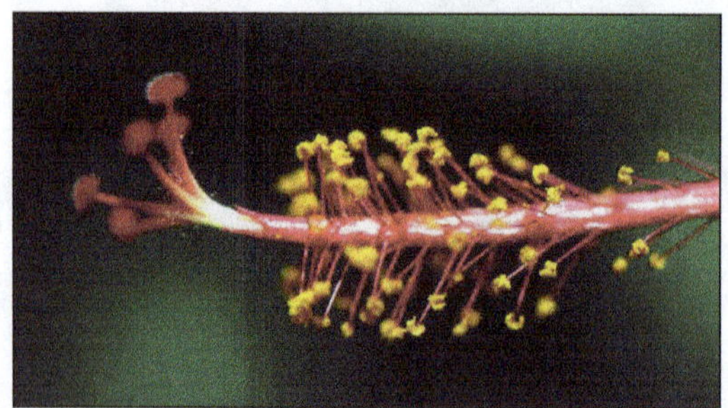

The pollen-bearing stamens (male structures) of a hibiscus flower, borne on the pistil (female structure). This unusual arrangement is common among members of the family Malvaceae.

A drooping male catkin (left) and the small red female inflorescence (right) of hazel (Corylus avellana).

Gynoecium

Gynoecium is most commonly used as a collective term for the parts of a flower that produce ovules and ultimately develop into the fruit and seeds. The gynoecium is the innermost whorl of a flower; it consists of (one or more) pistils and is typically surrounded by the pollen-producing reproductive organs, the stamens, collectively called the androecium. The gynoecium is often referred to as the "female" portion of the flower, although rather than directly producing female gametes (i.e. egg cells), the gynoecium produces megaspores, each of which develops into a female gametophyte which then produces egg cells.

The term gynoecium is also used by botanists to refer to a cluster of archegonia and any associated modified leaves or stems present on a gametophyte shoot in mosses, liverworts, and hornworts. The corresponding terms for the male parts of those plants are clusters of antheridia within the androecium. Flowers that bear a gynoecium but no stamens are called pistillate or carpellate. Flowers lacking a gynoecium are called staminate.

The gynoecium is often referred to as female because it gives rise to female (egg-producing) gametophytes; however, strictly speaking sporophytes do not have a sex, only gametophytes do. Gynoecium development and arrangement is important in systematic research and identification of angiosperms, but can be the most challenging of the floral parts to interpret.

Flower of Magnolia × wieseneri showing the many pistils making up the gynoecium in the middle of the flower.

Hippeastrum stigmas and style.

Pistils

The gynoecium may consist of one or more separate pistils. A pistil typically consists of an expanded basal portion called the ovary, an elongated section called a style and an apical structure that receives pollen called a stigma.

- The ovary, is the enlarged basal portion which contains placentas, ridges of tissue bearing one or more ovules (integumented megasporangia). The placentas and ovules may be born on the gynoecial appendages or less frequently on the floral apex. The chamber in which the ovules develop is called a locule (or sometimes cell).

- The style is a pillar-like stalk through which pollen tubes grow to reach the ovary. Some flowers such as Tulipa do not have a distinct style, and the stigma sits directly on the ovary. The style is a hollow tube in some plants such as lilies, or has transmitting tissue through which the pollen tubes grow.

- The stigma is usually found at the tip of the style, the portion of the carpels that receives pollen (male gametophytes). It is commonly sticky or feathery to capture pollen.

Carpels

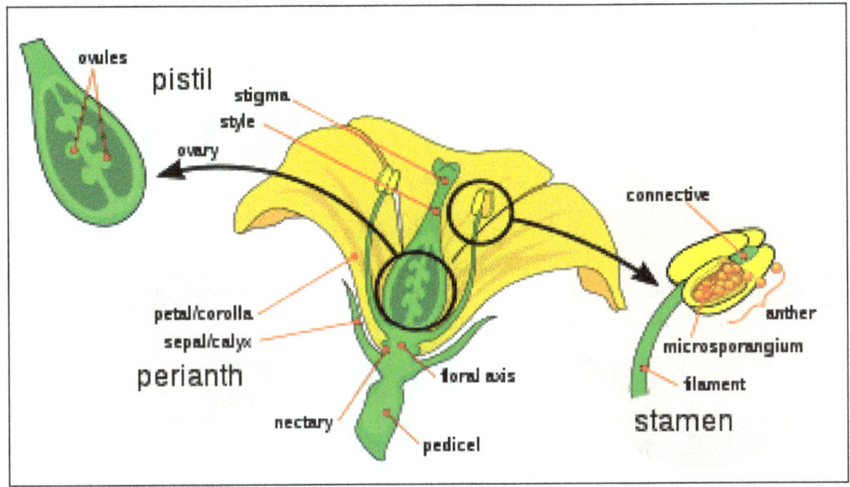

Mature flower.

The pistils of a flower are considered to be composed of carpels. A carpel is the female reproductive part of the flower, interpreted as modified leaves bearing structures called ovules, inside which the egg cells ultimately form and composed of ovary, style and stigma. A pistil may consist of one carpel, with its ovary, style and stigma, or several carpels may be joined together with a single ovary, the whole unit called a pistil. The gynoecium may consist of one or more uni-carpellate (with one carpel) pistils, or of one multi-carpellate pistil. The number of carpels is described by terms such as tricarpellate (three carpels).

Carpels are thought to be phylogenetically derived from ovule-bearing leaves or leaf homologues (megasporophylls), which evolved to form a closed structure containing the ovules. This structure is typically rolled and fused along the margin.

Although many flowers satisfy the above definition of a carpel, there are also flowers that do not have carpels according to this definition because in these flowers the ovules, although enclosed, are borne directly on the shoot apex. Different remedies have been suggested for this problem. An easy remedy that applies to most cases is to redefine the carpel as an appendage that encloses ovules and may or may not bear them.

Types

If a gynoecium has a single carpel, it is called monocarpous. If a gynoecium has multiple, distinct (free, unfused) carpels, it is apocarpous. If a gynoecium has multiple carpels "fused" into a single structure, it is syncarpous. A syncarpous gynoecium can sometimes appear very much like a monocarpous gynoecium.

Centre of a Ranunculus repens (creeping buttercup) showing multiple unfused carpels surrounded by longer stamens.

Cross-section through the ovary of Narcissus showing multiple connate carpels (a compound pistil) fused along the placental line where the ovules form in each locule.

Table: Comparison of gynoecium terminology using carpel and pistil.

Gynoecium composition	Terminology	Pistil terminology	Examples
Single carpel	Monocarpous (unicarpellate) gynoecium	A pistil (simple)	Avocado (Persea sp.), most legumes (Fabaceae)
Multiple distinct ("unfused") carpels	Apocarpous (choricarpous) gynoecium	Pistils (simple)	Strawberry (Fragaria sp.),Buttercup (Ranunculus sp.)
Multiple connate ("fused") carpels	Syncarpous gynoecium	A pistil (compound)	Tulip (Tulipa sp.), most flowers

The degree of connation ("fusion") in a syncarpous gynoecium can vary. The carpels may be "fused" only at their bases, but retain separate styles and stigmas. The carpels may be "fused" entirely, except for retaining separate stigmas. Sometimes (e.g., Apocynaceae) carpels are fused by their styles or stigmas but possess distinct ovaries. In a syncarpous gynoecium, the "fused" ovaries of the constituent carpels may be referred to collectively as a single compound ovary. It can be a challenge to determine how many carpels fused to form a syncarpous gynoecium+. If the styles and stigmas are distinct, they can usually be counted to determine the number of carpels. Within the compound ovary, the carpels may have distinct locules divided by walls called septa. If a syncarpous gynoecium has a single style and stigma and a single locule in the ovary, it may be necessary to examine how the ovules are attached. Each carpel will usually have a distinct line of placentation where the ovules are attached.

Pistil Development

Pistils begin as small primordia on a floral apical meristem, forming later than, and closer to the (floral) apex than sepal, petal and stamen primordia. Morphological and molecular studies of pistil ontogeny reveal that carpels are most likely homologous to leaves.

A carpel has a similar function to a megasporophyll, but typically includes a stigma, and is fused, with ovules enclosed in the enlarged lower portion, the ovary.

In some basal angiosperm lineages, Degeneriaceae and Winteraceae, a carpel begins as a shallow cup where the ovules develop with laminar placentation, on the upper surface of the carpel. The carpel eventually forms a folded, leaf-like structure, not fully sealed at its margins. No style exists, but a broad stigmatic crest along the margin allows pollen tubes access along the surface and between hairs at the margins.

Two kinds of fusion have been distinguished: postgenital fusion that can be observed during the development of flowers, and congenital fusion that cannot be observed i.e., fusions that occurred during phylogeny. But it is very difficult to distinguish fusion and non-fusion processes in the evolution of flowering plants. Some processes that have been considered congenital (phylogenetic) fusions appear to be non-fusion processes such as, for example, the de novo formation of intercalary growth in a ring zone at or below the base of primordia. Therefore, "it is now increasingly acknowledged that the term 'fusion,' as applied to phylogeny (as in 'congenital fusion') is ill-advised."

Gynoecium Position

Flowers and fruit (capsules) of the ground orchid,
Spathoglottis plicata, illustrating an inferior ovary.

Basal angiosperm groups tend to have carpels arranged spirally around a conical or dome-shaped receptacle. In later lineages, carpels tend to be in whorls.

The relationship of the other flower parts to the gynoecium can be an important systematic and taxonomic character. In some flowers, the stamens, petals, and sepals are often said to be "fused" into a "floral tube" or hypanthium. However, as Leins & Erbar pointed out, "the classical view that the wall of the inferior ovary results from the "congenital" fusion of dorsal carpel flanks and the floral axis does not correspond to the ontogenetic processes that can actually be observed. All that can be seen is an intercalary growth in a broad circular zone that changes the shape of the floral axis (receptacle)." And what happened during evolution is not a phylogenetic fusion but the formation of a unitary intercalary meristem. Evolutionary developmental biology investigates such developmental processes that arise or change during evolution.

If the hypanthium is absent, the flower is hypogynous, and the stamens, petals, and sepals are all attached to the receptacle below the gynoecium. Hypogynous flowers are often referred to as having a superior ovary. This is the typical arrangement in most flowers.

If the hypanthium is present up to the base of the styles, the flower is epigynous. In an epigynous flower, the stamens, petals, and sepals are attached to the hypanthium at the top of the ovary or, occasionally, the hypanthium may extend beyond the top of the ovary. Epigynous flowers are often referred to as having an inferior ovary. Plant families with epigynous flowers include orchids, asters, and evening primroses.

Between these two extremes are perigynous flowers, in which a hypanthium is present, but is either free from the gynoecium (in which case it may appear to be a cup or tube surrounding the gynoecium) or connected partly to the gynoecium (with the stamens, petals, and sepals attached to the hypanthium part of the way up the ovary). Perigynous flowers are often referred to as having a half-inferior ovary (or, sometimes, partially inferior or half-superior). This arrangement is particularly frequent in the rose family and saxifrages.

Occasionally, the gynoecium is born on a stalk, called the gynophore, as in *Isomeris arborea*.

Placentation

Within the ovary, each ovule is born by a placenta or arises as a continuation of the floral apex. The placentas often occur in distinct lines called lines of placentation. In monocarpous or apocarpous gynoecia, there is typically a single line of placentation in each ovary. In syncarpous gynoecia, the lines of placentation can be regularly spaced along the wall of the ovary (parietal placentation), or near the center of the ovary. In the latter case, separate terms are used depending on whether or not the ovary is divided into separate locules. If the ovary is divided, with the ovules born on a line of placentation at the inner angle of each locule, this is axile placentation. An ovary with free central placentation, on the other hand, consists of a single compartment without septae and the ovules are attached to a central column that arises directly from the floral apex (axis). In some cases a single ovule is attached to the bottom or top of the locule (basal or apical placentation, respectively).

The Ovule

Longitudinal section of carpellate flower of squash
showing ovary, ovules, stigma, style, and petals.

In flowering plants, the ovule is a complex structure born inside ovaries. The ovule initially consists of a stalked, integumented megasporangium (also called the nucellus). Typically, one cell in the megasporangium undergoes meiosis resulting in one to four megaspores. These develop into a megagametophyte (often called the embryo sac) within the ovule. The megagametophyte typically develops a small number of cells, including two special cells, an egg cell and a binucleate central cell, which are the gametes involved in double fertilization. The central cell, once fertilized by a sperm cell from the pollen becomes the first cell of the endosperm, and the egg cell once fertilized become the zygote that develops into the embryo. The gap in the integuments through which the pollen tube enters to deliver sperm to the egg is called the micropyle. The stalk attaching the ovule to the placenta is called the funiculus.

Role of the Stigma and Style

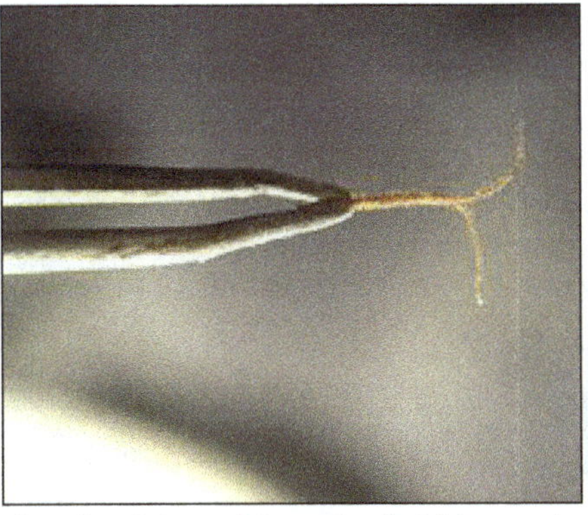

Stigmas and style of Cannabis sativa held in a pair of forceps.

Stigmas can vary from long and slender to globe-shaped to feathery. The stigma is the receptive tip of the carpels, which receives pollen at pollination and on which the pollen grain germinates. The stigma is adapted to catch and trap pollen, either by combining pollen of visiting insects or by various hairs, flaps, or sculpturings.

The style and stigma of the flower are involved in most types of self incompatibility reactions. Self-incompatibility, if present, prevents fertilization by pollen from the same plant or from genetically similar plants, and ensures outcrossing.

Stigma

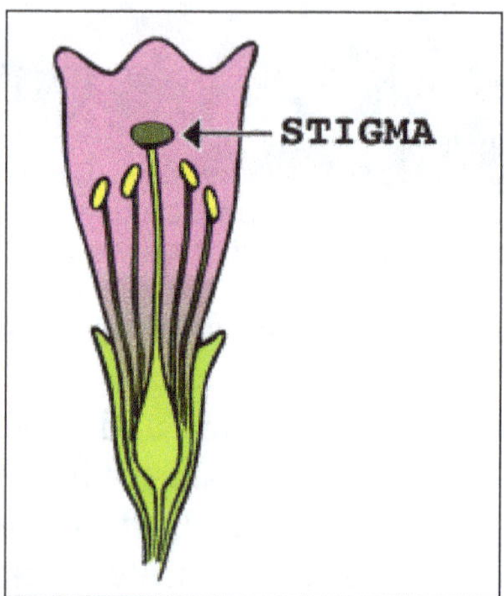

Diagram of stigma.

The stigma (plural: stigmas or stigmata) is the receptive tip of a carpel, or of several fused carpels, in the gynoecium of a flower.

The stigma, together with the style and ovary comprises the pistil, which in turn is part of the gynoecium or female reproductive organ of a plant. The stigma forms the distal portion of the style or stylodia. The stigma is composed of stigmatic papillae, the cells which are receptive to pollen. These may be restricted to the apex of the style or, especially in wind pollinated species, cover a wide surface.

Stigma of a Tulipa species, with pollen.

The stigma receives pollen and it is on the stigma that the pollen grain germinates. Often sticky, the stigma is adapted in various ways to catch and trap pollen with various hairs, flaps, or sculpturings. The pollen may be captured from the air (wind-borne pollen, anemophily), from visiting insects or other animals (biotic pollination), or in rare cases from surrounding water (hydrophily). Stigma can vary from long and slender to globe shaped to feathery.

Corn stigma called "silk".

Pollen is typically highly desiccated when it leaves an anther. Stigma have been shown to assist in the rehydration of pollen and in promoting germination of the pollen tube. Stigma also ensure proper adhesion of the correct species of pollen. Stigma can play an active role in pollen discrimination and some self-incompatibility reactions, that reject pollen from the same or genetically similar plants, involve interaction between the stigma and the surface of the pollen grain.

Shape

The stigma is often split into lobes, e.g. trifid (three lobed), and may resemble the head of a pin (capitate), or come to a point (punctiform). The shape of the stigma may vary considerably:

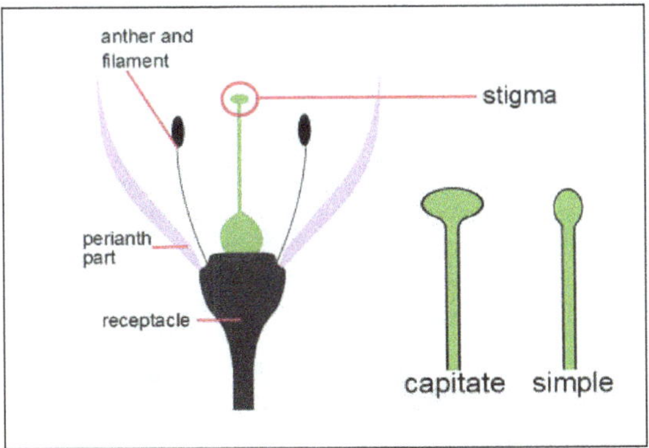

Capitate and simple.

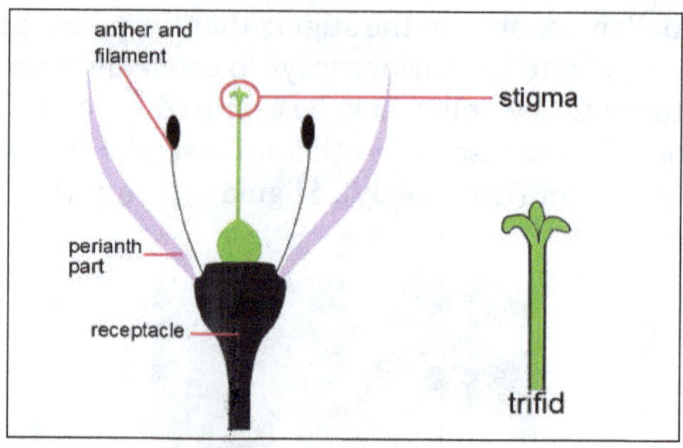

Trifid.

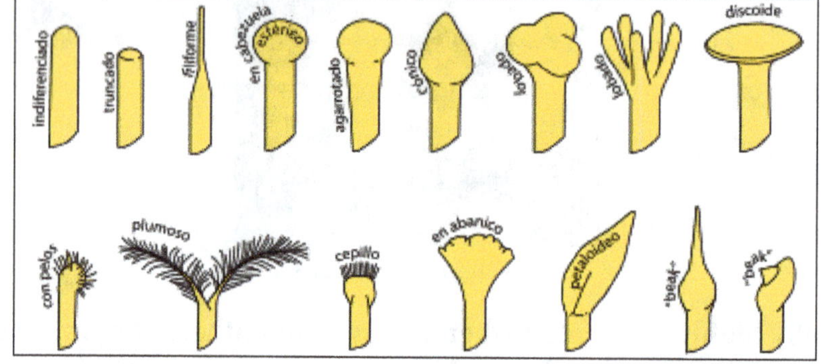

Style

Structure

The style is a narrow upward extension of the ovary, connecting it to the stigmatic papillae. It may be absent in some plants in the case the stigma is referred to as sessile. Styles are generally tube-like—either long or short. The style can be open (containing few or no cells in the central portion) with a central canal which may be filled with mucilage. Alternatively the style may be closed (densely packed with cells throughout). Most syncarpous monocots and some eudicots have open styles, while many syncarpous eudicots and grasses have closed (solid) styles containing specialised secretory transmitting tissue, linking the stigma to the centre of the ovary. This forms a nutrient rich tract for pollen tube growth.

Iris versicolor showing three structures with two overlapping lips, an upper petaloid style branch and a lower tepal, enclosing a stamen.

Where there are more than one carpel to the pistil, each may have a separate style-like stylodium, or share a common style. In Irises and others in the family Iridaceae, the style divides into three petal-like (petaloid) style branches (sometimes also referred to as 'stylodia'), almost to the base of the style and is called tribrachiate. These are flaps of tissue, running from the perianth tube above the sepal. The stigma is a rim or edge on the underside of the branch, near the end lobes. Style branches also appear on Dietes, Pardanthopsis and most species of Moraea.

The feathery stigma of Crocus speciosus has branches corresponding to three carpels.

In Crocuses, there are three divided style branches, creating a tube. Hesperantha has a spreading style branch. Alternatively the style may be lobed rather than branched. Gladiolus has a bi-lobed style branch (bilobate). Freesia, Lapeirousia, Romulea, Savannosiphon and Watsonia have bifuracated (two branched) and recurved style branches.

Attachment to the Ovary

May be terminal (apical), subapical, lateral, gynobasic, or subgynobasic. Terminal (apical) style position refers to attachment at the apex of the ovary and is the commonest pattern. In the subapical pattern the style arises to the side slightly below the apex. a lateral style arises from the side of the ovary and is found in Rosaceae. The gynobasic style arises from the base of the ovary, or between the ovary lobes and is characteristic of Boraginaceae. Subgynobasic styles characterise Allium.

Style postion.

Termianl (apical).

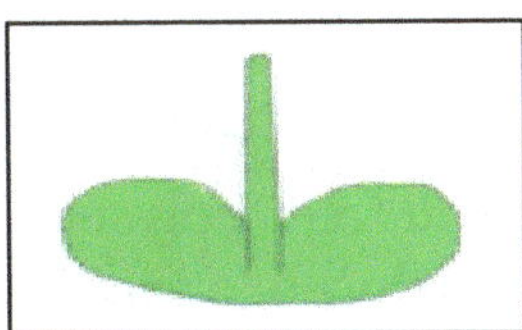

Gynobasic.

Pollination

Pollen tubes grow the length of the style to reach the ovules, and in some cases self-incompatibility reactions in the style prevent full growth of the pollen tubes. In some species, including Gasteria at least, the pollen tube is directed to the micropyle of the ovule by the style.

Sepal

A sepal is a part of the flower of angiosperms (flowering plants). Usually green, sepals typically function as protection for the flower in bud, and often as support for the petals when in bloom. The term sepalum was coined by Noël Martin Joseph de Necker in 1790. Collectively the sepals are called the calyx (plural calyces), the outermost whorl of parts that form a flower.

After flowering, most plants have no more use for the calyx which withers or becomes vestigial. Some plants retain a thorny calyx, either dried or live, as protection for the fruit or seeds. Examples include species of Acaena, some of the Solanaceae (for example the Tomatillo, Physalis philadelphica), and the water caltrop, Trapa natans. In some species the calyx not only persists after flowering, but instead of withering, begins to grow until it forms a bladder-like enclosure around the fruit. This is an effective protection against some kinds of birds and insects, for example in Hibiscus trionum and the Cape gooseberry. In other species, the calyx grows into an accessory fruit.

Morphologically, both sepals and petals are modified leaves. The calyx (the sepals) and the corolla (the petals) are the outer sterile whorls of the flower, which together form what is known as the perianth.

The term tepal is usually applied when the parts of the perianth are difficult to distinguish, e.g. the petals and sepals share the same color, or the petals are absent and the sepals are colorful. When the undifferentiated tepals resemble petals, they are referred to as "petaloid", as in petaloid monocots, orders of monocots with brightly coloured tepals. Since they include Liliales, an alternative name is lilioid monocots. Examples of plants in which the term tepal is appropriate include genera such as Aloe and Tulipa. In contrast, genera such as Rosa and Phaseolus have well-distinguished sepals and petals.

The number of sepals in a flower is its merosity. Flower merosity is indicative of a plant's classification. The merosity of a eudicot flower is typically four or five. The merosity of a monocot or palaeodicot flower is three, or a multiple of three.

The development and form of the sepals vary considerably among flowering plants. They may be free (polysepalous) or fused together (gamosepalous). Often, the sepals are much reduced, appearing somewhat awn-like, or as scales, teeth, or ridges. Most often such structures protrude until the fruit is mature and falls off.

Examples of flowers with much reduced perianths are found among the grasses.

In some flowers, the sepals are fused towards the base, forming a calyx tube (as in the Lythraceae family, and Fabaceae). In other flowers (e.g., Rosaceae, Myrtaceae) a hypanthium includes the bases of sepals, petals, and the attachment points of the stamens.

Petal

Petals are modified leaves that surround the reproductive parts of flowers. They are often brightly colored or unusually shaped to attract pollinators. Together, all of the petals of a flower are called a corolla. Petals are usually accompanied by another set of special leaves called sepals, that collectively form the calyx and lie just beneath the corolla. The calyx and the corolla together make up the perianth. When the petals and sepals of a flower are difficult to distinguish, they are collectively called tepals. Examples of plants in which the term tepal is appropriate include genera such as Aloe and Tulipa. Conversely, genera such as Rosa and Phaseolus have well-distinguished sepals and petals. When the undifferentiated tepals resemble petals, they are referred to as "petaloid", as in petaloid monocots, orders of monocots with brightly coloured tepals. Since they include Liliales, an alternative name is lilioid monocots.

Although petals are usually the most conspicuous parts of animal-pollinated flowers, wind-pollinated species, such as the grasses, either have very small petals or lack them entirely.

Tetrameric flower of a Primrose willowherb (Ludwigia octovalvis) showing petals and sepals.

A Tulip's actinomorphic flower with three petals and three sepals, that collectively present a good example of an undifferentiated perianth. In this case, the word "tepals" is used.

Corolla

Apopetalous corolla.

The role of the corolla in plant evolution has been studied extensively since Charles Darwin postulated a theory of the origin of elongated corollae and corolla tubes.

Tubular-campanulate corolla, bearing long points and emergent from tubular calyx (Brugmansia aurea, Golden Angel's Trumpet, family Solanaceae).

A corolla of separate tepals, without fusion of individual segments, is apopetalous. If the petals are free from one another in the corolla, the plant is polypetalous or choripetalous; while if the petals are at least partially fused together, it is gamopetalous or sympetalous. In the case of fused tepals, the term is syntepalous. The corolla in some plants forms a tube.

Variations

Petals can differ dramatically in different species. The number of petals in a flower may hold clues to a plant's classification. For example, flowers on eudicots (the largest group of dicots) most frequently have four or five petals while flowers on monocots have three or six petals, although there are many exceptions to this rule.

The petal whorl or corolla may be either radially or bilaterally symmetrical. If all of the petals are essentially identical in size and shape, the flower is said to be regular or actinomorphic (meaning "ray-formed"). Many flowers are symmetrical in only one plane (i.e., symmetry is bilateral) and are termed irregular or zygomorphic (meaning "yoke-"

or "pair-formed"). In irregular flowers, other floral parts may be modified from the regular form, but the petals show the greatest deviation from radial symmetry. Examples of zygomorphic flowers may be seen in orchids and members of the pea family.

In many plants of the aster family such as the sunflower, Helianthus annuus, the circumference of the flower head is composed of ray florets. Each ray floret is anatomically an individual flower with a single large petal. Florets in the centre of the disc typically have no or very reduced petals. In some plants such as Narcissus the lower part of the petals or tepals are fused to form a floral cup (hypanthium) above the ovary, and from which the petals proper extend.

Petal often consists of two parts: the upper, broad part, similar to leaf blade, also called the blade and the lower part, narrow, similar to leaf petiole, called the claw, separated from each other at the limb. Claws are developed in petals of some flowers of the family Brassicaceae, such as Erysimum cheiri.

The inception and further development of petals shows a great variety of patterns. Petals of different species of plants vary greatly in colour or colour pattern, both in visible light and in ultraviolet. Such patterns often function as guides to pollinators, and are variously known as nectar guides, pollen guides, and floral guides.

Genetics

The genetics behind the formation of petals, in accordance with the ABC model of flower development, are that sepals, petals, stamens, and carpels are modified versions of each other. It appears that the mechanisms to form petals evolved very few times (perhaps only once), rather than evolving repeatedly from stamens.

Significance of Pollination

Pollination is an important step in the sexual reproduction of higher plants. Pollen is produced by the male flower or by the male organs of hermaphroditic flowers.

Pollen does not move on its own and thus requires wind or animal pollinators to disperse the pollen to the stigma (botany) of the same or nearby flowers. However, pollinators are rather selective in determining the flowers they choose to pollinate. This develops competition between flowers and as a result flowers must provide incentives to appeal to pollinators (unless the flower self-pollinates or is involved in wind pollination). Petals play a major role in competing to attract pollinators. Henceforth pollination dispersal could occur and the survival of many species of flowers could prolong.

Functions and Purposes

Petals have various functions and purposes depending on the type of plant. In general, petals operate to protect some parts of the flower and attract/repel specific pollinators.

Function

This is where the positioning of the flower petals are located on the flower is the corolla e.g. the buttercup having shiny yellow flower petals which contain guidelines amongst the petals in aiding the pollinator towards the nectar. Pollinators have the ability to determine specific flowers they wish to pollinate. Using incentives flowers draw pollinators and set up a mutual relation between each other in which case the pollinators will remember to always guard and pollinate these flowers (unless incentives are not consistently met and competition prevails).

Scent

The petals could produce different scents to allure desirable pollinators or repel undesirable pollinators. Some flowers will also mimic the scents produced by materials such as decaying meat, to attract pollinators to them.

Colour

Various colour traits are used by different petals that could attract pollinators that have poor smelling abilities, or that o1nly come out at certain parts of the day. Some flowers are able to change the colour of their petals as a signal to mutual pollinators to approach or keep away.

Shape and Size

Furthermore, the shape and size of the flower/petals is important in selecting the type of pollinators they need. For example, large petals and flowers will attract pollinators at a large distance or that are large themselves. Collectively the scent, colour and shape of petals all play a role in attracting/repelling specific pollinators and providing suitable conditions for pollinating. Some pollinators include insects, birds, bats and the wind. In some petals, a distinction can be made between a lower narrowed, stalk-like basal part referred to as the claw, and a wider distal part referred to as the blade (or limb). Often the claw and blade are at an angle with one another.

Types of Pollination

Wind Pollination

Wind-pollinated flowers often have small, dull petals and produce little or no scent. Some of these flowers will often have no petals at all. Flowers that depend on wind pollination will produce large amounts of pollen because most of the pollen scattered by the wind tends to not reach other flowers.

Attracting Insects

Flowers have various regulatory mechanisms in order to attract insects. One such helpful mechanism is the use of colour guiding marks. Insects such as the bee or butterfly can see the ultraviolet marks which are contained on these flowers, acting as an attractive mechanism which is not visible towards the human eye. Many flowers contain a variety of shapes acting to aid with the landing of the visiting insect and also influence the insect to brush against anthers and stigmas (parts of the flower). One such example of a flower is the pohutukawa (Metrosideros excelsa) which acts to regulate colour within a different way. The pohutukawa contains small petals also having bright large red clusters of stamens. Another attractive mechanism for flowers is the use of scents which are highly attractive to humans. One such example is the rose. On the other hand, some flowers produce the smell of rotting meat and are attractive to insects such as flies. Darkness is another factor which flowers have adapted to as nighttime conditions limit vision and color-perception. Fragrancy can be especially useful for flowers which are pollinated at night by moths and other flying insects.

Attracting Birds

Flowers are also pollinated by birds and must be large and colorful to be visible against natural scenery. In New Zealand, such bird–pollinated native plants include: kowhai (Sophora species), flax (Phormium tenax) and kaka beak (Clianthus puniceus). Flowers adapt the mechanism on their petals to change colour in acting as a communicative mechanism for the bird to visit. An example is the tree fuchsia (Fuchsia excorticata) which are green when needing to be pollinated and turn red for the birds to stop coming and pollinating the flower.

Bat-pollinated Flowers

Flowers can be pollinated by short tailed bats. An example of this is the dactylanthus (Dactylanthus taylorii). This plant has its home under the ground acting the role of a parasite on the roots of forest trees. The dactylanthus has only its flowers pointing to the surface and the flowers lack colour but have the advantage of containing lots of nectar and a very strong scent. These act as a very useful mechanism in attracting the bat.

Perianth

Perianth (perigonium, perigon or perigone) is the non-reproductive part of the flower, and structure that forms an envelope surrounding the sexual organs, consisting of the calyx (sepals) and the corolla (petals). In the mosses and liverworts (Marchantiophyta), the perianth is the sterile tubelike tissue that surrounds the female reproductive structure (or developing sporophyte).

Flowering Plants

In flowering plants, the perianth may be described as being either dichlamydeous/heterochlamydeous in which the calyx and corolla are clearly separate, or homochlamydeous, in which they are indistinguishable (and the sepals and petals are collectively referred to as tepals). When the perianth is in two whorls, it is described as biseriate. While the calyx may be green, known as sepaloid, it may also be brightly coloured, and is then described as petaloid. When the undifferentiated tepals resemble petals, they are also referred to as "petaloid", as in petaloid monocots, orders of monocots with brightly coloured tepals. Since they include Liliales, an alternative name is lilioid monocots. The corolla and petals have a role in attracting pollinators, but this may be augmented by more specialised structures like the corona.

When the corolla consists of separate tepals the term apotepalous is used, or syntepalous if the tepals are fused to one another. The petals may be united to form a tubular corolla (gamopetalous or sympetalous). If either the petals or sepals are entirely absent, the perianth can be described as being monochlamydeous.

References

- Flower-anatomy: proflowers.com, Retrieved 18 May, 2020

- Flower: cropsreview.com, Retrieved 25 January, 2020

- Judd, W.S.; Campbell, C.S.; Kellogg, E.A.; Stevens, P.F. & Donoghue, M.J. (2007). Plant Systematics: A Phylogenetic Approach (3rd ed.). Sunderland, MA: Sinauer Associates, Inc. ISBN 978-0-87893-407-2

- Stamen, science: britannica.com, Retrieved 17 February, 2020

- Graham, S. W.; Barrett, S. C. H. (1 July 2004). "Phylogenetic reconstruction of the evolution of stylar polymorphisms in Narcissus (Amaryllidaceae)". American Journal of Botany. 91 (7): 1007–1021. doi:10.3732/ajb.91.7.1007. PMID 21653457. Retrieved 25 October 2014

- Leaf: biologydictionary.net, Retrieved 28 August, 2020

- Simpson, Michael G. (2011). Plant Systematics. Academic Press. ISBN 0-08-051404-9. Retrieved 12 February 2014

Fruit and Seed Anatomy   5

Fruits are the mature ovaries of one or more flowers. Fruit anatomy is the internal structure of fruits. Seeds are matured and fertilized ovules. It consists of three components – embryo, endosperm and seed coat. This chapter closely examines fruit and seed anatomy to provide an extensive understanding of the subject.

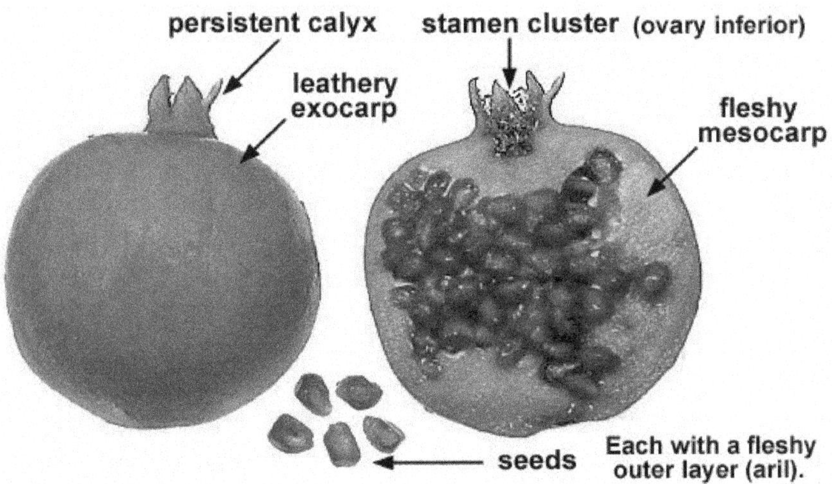

Pomegranate (*Punica granatum*): A many-seeded berry.

Fruit Anatomy

A fruit develops from a ripe ovary or any floral parts on the basis of floral parts they develop, fruits may be true or false.

True Fruits

A true fruit or eucarp is a mature or ripened ovary, developed after fertilization, e.g., Mango, Maize, Grape etc.

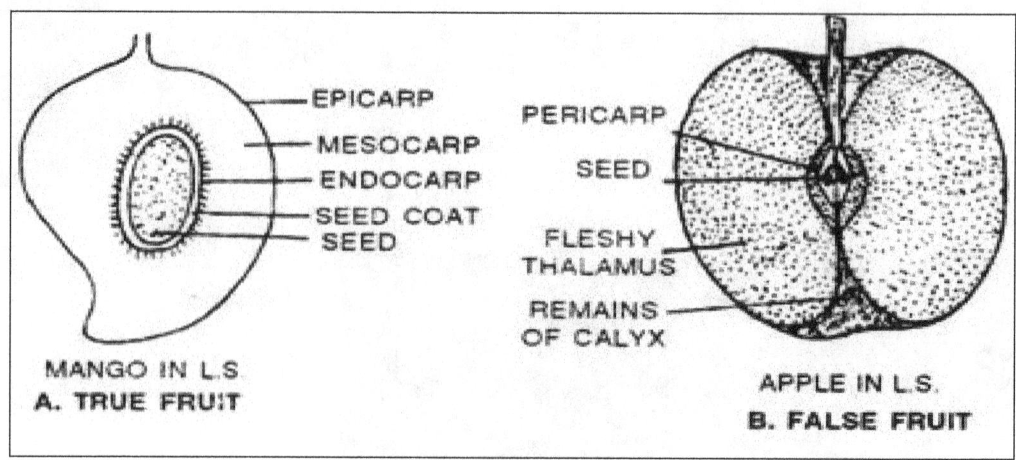

Morphology of true and false froits.

False Fruits

A false fruit or pseudo-carp is derived from the floral parts other than ovary, e.g., peduncle in cashew-nut, thalamus in apple, pear, gourd and cucumber; fused perianth in mulberry and calyx in Dillenia (Or. Ou). Jack fruit and pine apple are also false fruits as they develop from the entire inflorescence. False fruits are also called spurious or accessory fruits.

Parthenocarpic Fruits

These are seedless fruits that are formed without fertilization, e.g., Banana. Now a day many seedless grapes, oranges and water melones are being developed by horticulturists. Pomology is a branch of horticulture that deals with the study of fruits and their cultivation.

Morphology of a Typical Fruit

A fruit consists of pericarp and seeds. Seeds are fertilized and ripened ovules. The pericarp develops from the ovary wall and may be dry or fleshy. When fleshy, pericarp is differentiated into outer epicarp, middle mesocarp and inner endocarp.

Types of Fruits

On the basis of the above mentioned features, fruits are usually classified into three main groups:

- Simple.

- Aggregate.

- Composite or Multiple fruits.

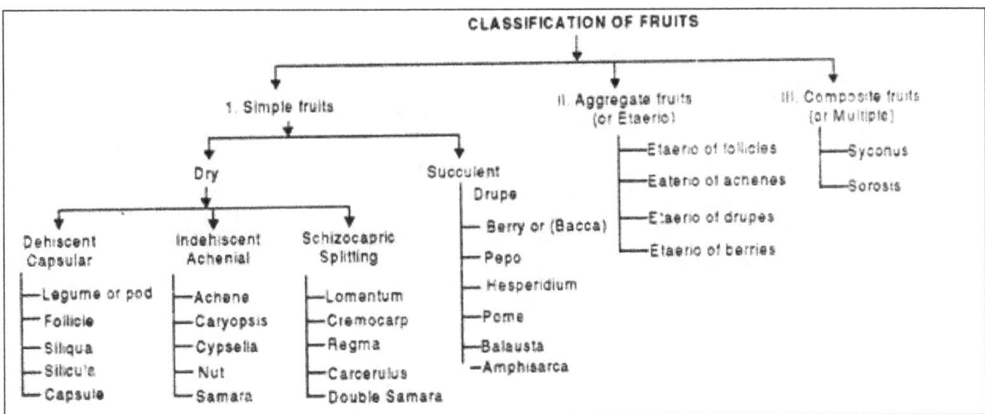

Simple Fruits

When a single fruit develops from a single ovary of a single flower, it is called a simple fruit. The ovary may belong to a monocarpellary simple gynoecium or to a polycarpellary syncarpous gynoecium. There are two categories of simple fruits—dry and fleshy.

Simple fruits are of two types:

- Dry Fruits: These fruits are not fleshy, and their pericarp (fruit wall) is not distinguished into three layers.

- Succulent Fruits (Fleshy fruits): In these fruits pericarp is distinguished into epicarp, mesocarp and endocarp. Mesocarp is fleshy or fibrous. These fruits are indehiscent, and seeds are liberated after the decay of the flesh.

Dry Fruits

Three types of dry fruits are distinguishable:

- Dehiscent Fruits (Capsular Fruits): Characteristic of these fruits is that their pericarp rupture after ripening and the seeds are disseminated.

- Indehiscent Fruits (Achenial Fruits): As their name indicates, pericarp of such fruits does not rupture on ripening and the seeds remain inside.

- Schizocarpic Fruits (Splitting Fruits): These fruits fall in between the above-mentioned two categories. Here, the fruit on ripening divides into one-seeded segments or mericarp; but the mericarps remain un-ruptured.

Dehiscent Fruits (Capsular Fruits)

Depending on the mode of dehiscence, these fruits can be divided into the following five classes:

- Legume or Pod: Legume develops from a superior, monocarpellary, unilocular ovary. At maturity, the fruit dehisces along both the sutures i.e. ventral as well as dorsal. It is characteristic of family Leguminosae (Pea, Gram etc).

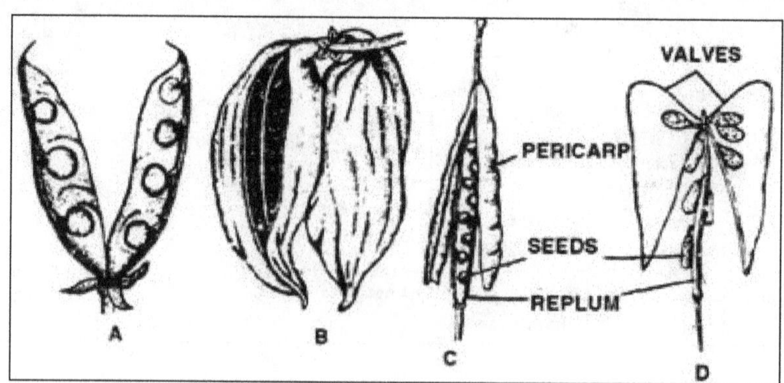

Dehiscent fruits A. Legume of Pea B. Follicles of Calatropis. C. Siliqua of Brasicca D. Silicula of Capsella.

- Follicle: It is similar to legume but it dehisces only along the ventral suture, e.g. Larkspur, Calatropis, Michelia, Vinca.

- Siliqua: Siliqua develops from a bicarpellary, syncarpous, superior ovary which is unilocular but becomes bilocular due to a false septum called replum. It is an elongated fruit in which dehiscence occurs along both the sutures from base to apex and the seeds attached to the replum get exposed. Example-Brassica (Mustard).

- Silicula: A short and flattened siliqua is called silicula. It is almost as broad as long. Examples: Iberis amara (Candytuft), Capsella bursa-pastoris (Shepherd's purse).

- Capsule: It is a simple dry many seeded dehiscent fruit developing from a multi-carpellary syncarpous ovary.

Types of Dehiscence Capsules

- Porocidal: The dehiscence occurs through pores as in Poppy (Papaver).

- Pyxis: This is a special name given to a capsule when the dehiscence is transverse so that the top comes off as a lid as if exposing a box of seeds, e.g., Celosia (Cock's comb), Amaranth us, Chalfweed.

- Loculicidal: The dehiscence occurs by longitudinal slits which open into the loculi, e.g., 1 dy's finger (Abelmoschus).

- Septicidal: The dehiscence line appears along the septa, e.g.. Linseed, Cotton to the central axis,eg. Datura.

- Sentifragai: Tin- broken parts separate exposing the seeds attached to the central axis, e.g Datum.

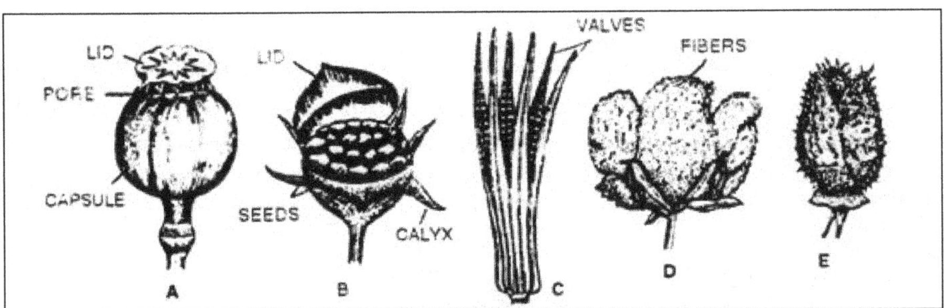

Capsule fruits A Porocidal capsule of poppy B. Pyxis of celosia C. Loculicidal capsule of abelmoschus, D. Septicidal Capsule of Gossypium E. Septifragal capsule of Datura.

Indehiscent or Achenial Fruits

Achenial fruits are simple, indehiscent, single seeded having a thin, dry, woody or leathery pericarp.

There are five common types of achenial fruits:

- Achene: The pericarp of the fruit is free from the testa of the seed. The seed is attached to the pericarp only at one point. It develops from superior monocarpellary pistill having unilocular and uniovuled ovary, e.g., Mirabilis jalapa, but more commonly achenes occur in the form of aggregate fruits as in Ranunculus and Clematis etc.

- Caryopsis: It is similar to achene except that in this case pericarp and testa are inseparably fused as in cereals. It is a characteristic future of family Gramineae. Example—Wheat, Maize etc.

- Cypsela: 11 is a characteristic feature of family Compositae. The fruit wall is free from testa and a typical feature of the fruit is the presence of a pappus having a crown of hair like processes which helps in wind-dispersal. The fruit develops from bicarpellary, syncarpous, interior ovary having a single basal ovule, e.g., Sonchus, Dandelion etc.

- Samara: It develops from a monocarpellary pistil with a superior, unilocular and uniovuled ovary. The pericarp is expanded in the form of wings which help in dispersal. Example—Holoptelea and Elm.

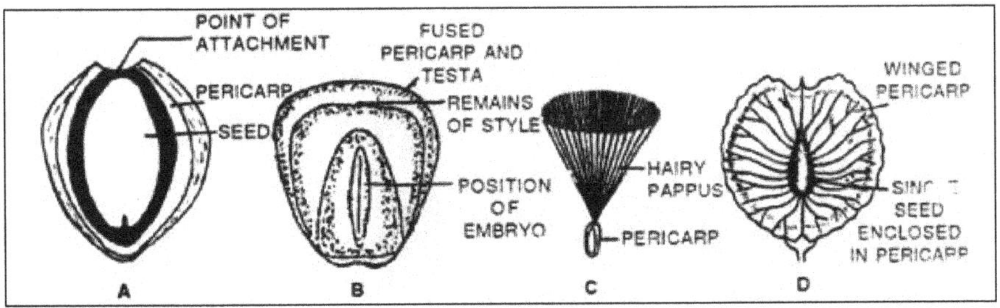

Achenial fruits A. Achene of Mirabilis (L.S) B. Caryopsis of Maize C. Crysele of Sonchus D. Samara of Elm.

- Nut: The pericarp is harder and leathery or woody. It may develop from a simple or compound pistil with superior or inferior, uniovuled ovary. Examples— Quercus (Oak), Litchi and Cashew-nut Trapa etc. In case of Litchi pericarp is hard and leathery. The edible part is aril which is an outgrowth of testa from the micropylar end and becomes juicy.

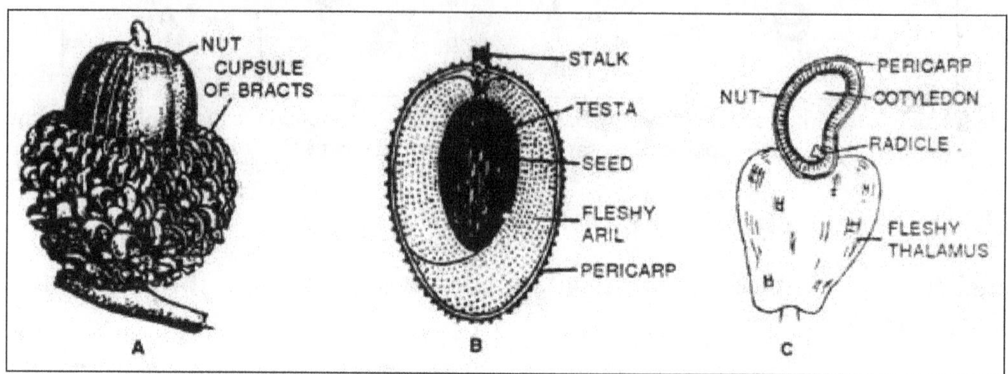

Nuts A. Nut of OAK B. Nut of Litchi C. Nut of Cashewnut.

Schizocarpic or Splitting Fruits

These fruits maybe considered intermediate between achenial (being indehiscent) and capsular (being many seeded) fruits. The fruit breaks up into a number of indehiscent single-seeded segments called mericarps from which seeds are liberated only when pericarp gets rotten. In some cases one- seeded parts of the fruit are dehiscent and are called Cocci.

Schizocarpic fruits are of following 5 types:

- Lomentum: The fruit is constricted between the seeds and usually breaks up into segments containing one or more seeds, e.g. Mimosa, Acacia arabica. In case of radish, the fruit is lomentaceous siliqua.

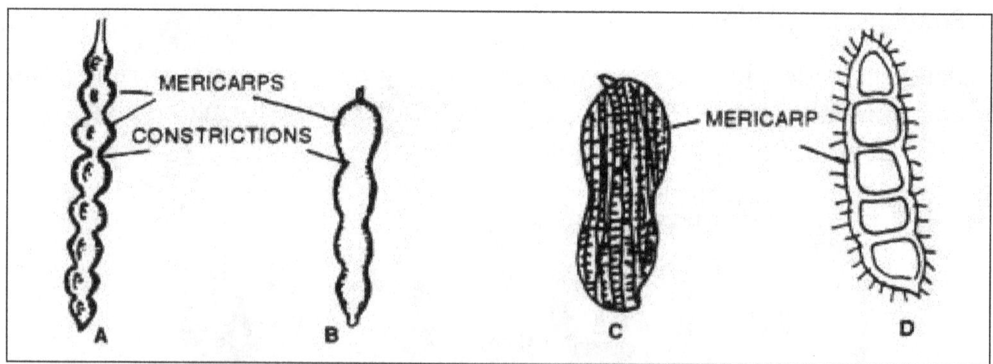

Lomentums A. Acacia B. Radish C. Groundnut D. Mimosa.

- Compound Samara: This is a type of two or more-chambered fruit derived from a syncarpous (i.e., compound) ovary. The pericarp is extended in the form of wings and at maturity the fruit breaks up % into single seeded mericarps. e.g., EIm (Holoptelea), maple.

- Cremocarp: This is a two-seeded fruit derived from bicarpellary, syncarpous, inferior, bilocular and uniovuled ovary. It is a typical fruit of family umbelliferae. The two mericarps split along the central axis or carpophore to which they remain attached. Persistent style and stylopodium are present e.g. Coriander.

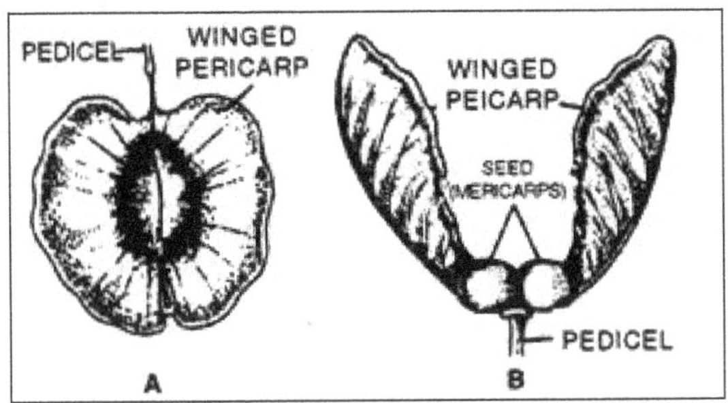

Compound Samara A. Holoptelea B. Acer.

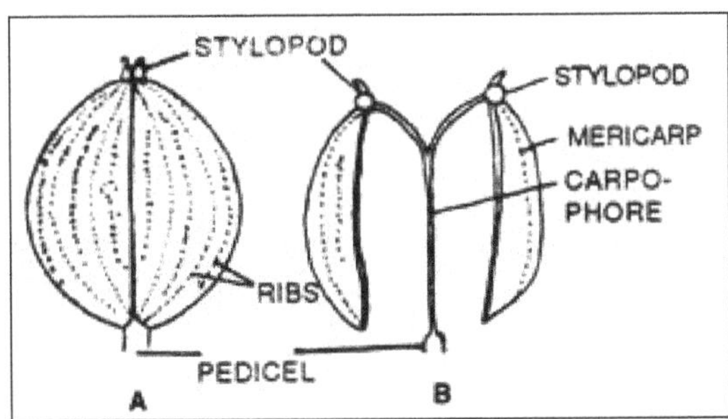

Cremocarp of Coriander A. Complete B. Mericarps Separated.

- Compound Samara: This is a type of two or more-chambered fruit derived from a syncarpous (i.e., compound) ovary. The pericarp is extended in the form of wings and at maturity the fruit breaks up % into single seeded mericarps. e.g., EIm (Holoptelea), maple.

- Carcerulas: This fruit is derived from superior, syncarpous pistil, multilocular with axile placentation. The fruit splits into many mericarps. e.g. Hollyhock (Althaea rosea), Salvia, Ocimum.

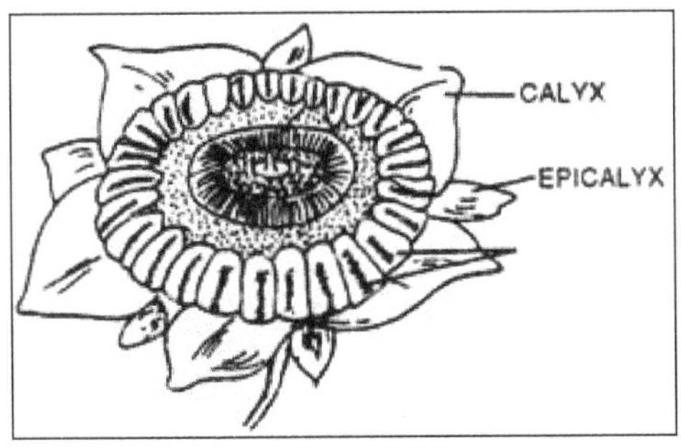

Carcerulus of althaea.

- **Regma:** It is derived from polycarpellary pistil which splits into as many Cocci (dehiscent segments) as there are carpels. Regma of castor breaks up into three cocci as it is derived from tricarpellary syncarpous pistil. Similarly, regma of Geranium breaks into five cocci as it is derived from five carpels.

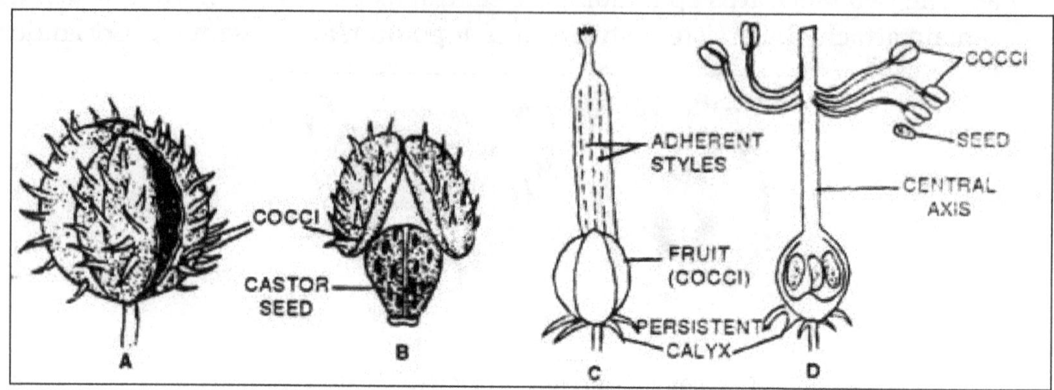

Regma A. Regma of castor breaking into three cocci, B. Coccus releasing a seed C. Regma of Geranium D. Dehisced regma of germanium.

Succulent or Fleshy Fruits

These are simple fruits with fleshy pericarp. The simple succulent fruits are of 3 types – drupe, pome and berrie.

- **Drupe:** The pericarp or fruit wall is differentiated into thin epicarp (skin) fleshy mesocarp and stony endocarp.Hence.it is also called as stone fruit, e.g., Mango, Coconut, Peach, Almond, Trapa etc. In mango, mesocarp is juicy and edible. In coconut mesocarp is fibrous and edible part is endocarp. In almond, epicarp and mesocarp get peeled off and only hard endocarp can be seen in marketed fruits. The edible part is cotyledons.

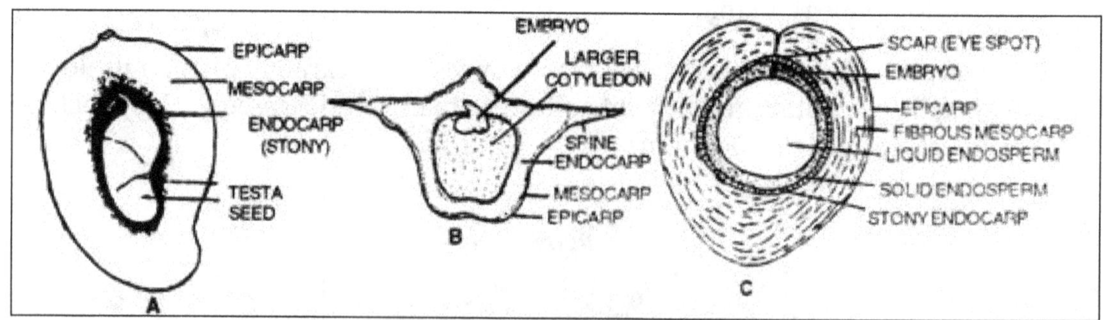

Drupes in: A- Mango, B- Trape, C- Coconut.

- **Pome:** It is a simple, fleshy but false fruit as it is surrounded by a fleshy thalamus which is edible while actual fruit lies within, e.g., apple, pear, loquat etc.

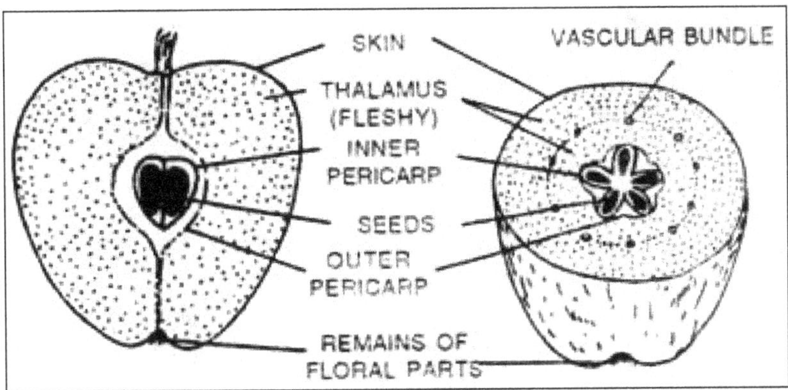

Pome of Apple.

- **Berry and Bacca:** Berry is a fleshy fruit in which there is no hard part except the seeds. Pericarp may be differentiated into epicarp, mesocarp and endocarp. One or other of these layers may form pulp in which seeds are embedded which generally gets detached from the placenta.

  The fruits derived from superior ovary are called superior or true berries as in brinjal, grape, tomato. False berries are derived from inferior ovary and thalamus and pericarp are fused as in banana and guava etc. In case of epicarp and thalamus are peeled off, mesocarp and endocarp with embedded unripe seeds forms the edible part. In case of Date, epicarp and mesocarp are edible while papery and thin endocarp is thrown away along with the seed.

  There are some fruits which show variations from the normal berry:

  - **Pepo:** This develops from inferior ovary which is unilocular or falsely trilocular having parietal placentation. The seeds remain attached to placenta. The outer ring is very hard as in Cucurbits.

  - **Hesperidium:** It develops from polycarpellary, syncarpous, superior, multilocuiar ovary with axile placentation. Epicarp forms the leathery peeling, mesocarp is in the form of fibres while the endocarp projects inwards forming distinct chambers from which juicy ingrowths in the form of hair arise which form the edible part, eg. Citrus (Orange, Lemon).

  - **Amphisarca:** It is derived from polycarpellary, syncarpous, multilocuiar and superior ovary. In this case, epicarp is woody. The placenta and inner layers of pericarp become pulpy and edible in which the seeds are scattered. The testa is muclilagenous, e.g., Aegle marmelose.

  - **Balusta:** It is a berry with an outer hard rind formed of epicarp and a part of mesocarp. The inward foldings of mesocarp form chambers. Each chamber is lined by papery endocarp which encloses a group of seeds. The seeds are covered by edible juicy testa. e.g., Pomranate.

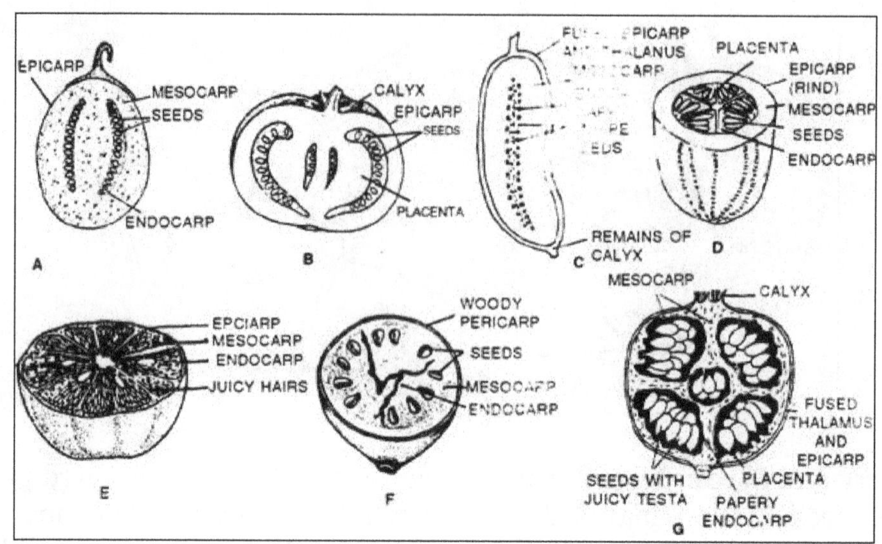

Berries of baccate fruits. A: Berry of Bengal, B: Berry of Tomato, C: Berry of Banana, D: Pepo of Cucumber, E: Hesperidium of Lemon, F: Amphisaraca of Wood Apple, G: Balusta of Pomengranate.

Aggregate Fruits

Flowers with polycarpellary and apocarpous gynoecium give rise to a number of fruitlets as there are a number of free ovaries, each giving rise to one fruitlet. Sometimes, these fruitlets coalesce together appearing to be a single fruit but in many other cases, the fruitlets remain free from one another forming etaerio of fruitlets. An aggregate fruit is named according to the nature of fruitlets.

- Etaerio of achenes: Aggregate of achenes are found in Fragaria (strawberry), Rose, Ranunculus, Nelumbium (lotus) etc. Here each fruitlet is an achene; and achenes are hairy. In rose (Rosa), many achenes are present on a saucer (cup) – shaped thalamus. In lotus (Nelumbium), thalamus becomes spongy and some achenes are embedded in it. In strawberry [Fragaria), the thalamus is fleshy and becomes red on maturation and is the edible part.

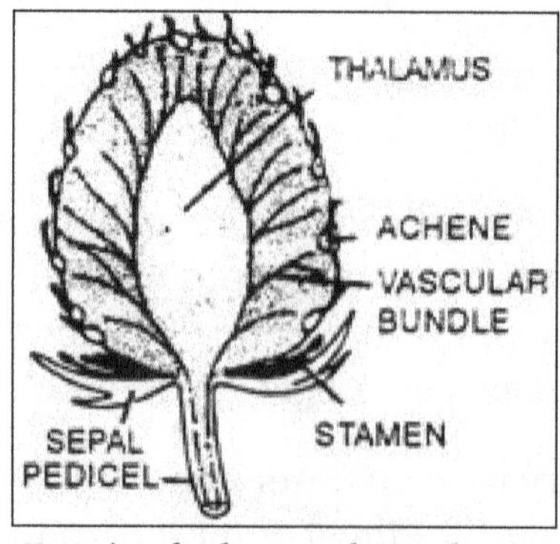

Etaerio of achenes of strawberry.

- Etaerio of follicles: Etaerio of follicles can be seen in Aconitum, Catotropis,

Crypiostegia etc. In Aconitum three fruitlets from each flower while two fruitlets (follicles) develop from one flower in Calotroiis, Cryptostegia and Michelia.

Etaerio of follicles of Michelia.

- Etaerio of samaras: It can be studied in Ailanthus where many winged samaras develop from one flower.

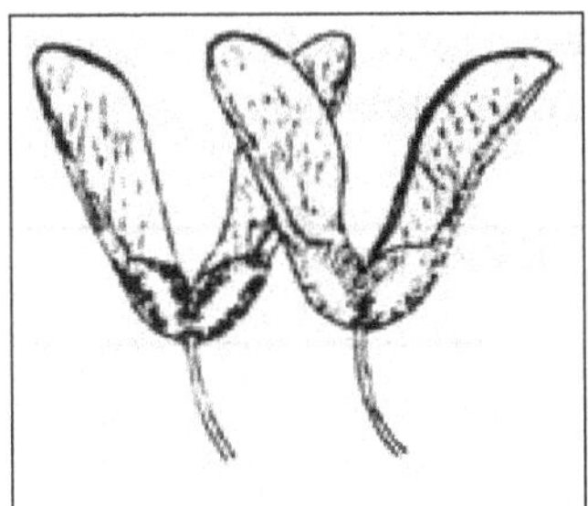

Etaerio of samaras of Alianthus.

- Etaerio of berries: In Artabotrys berries occur in a bunch. In Anona squomosa (Custard apple) the berries become very fleshy and being crowded together on a thick thalamus form a complex single fruit. The apices of berries fuse together forming something like a common rind.

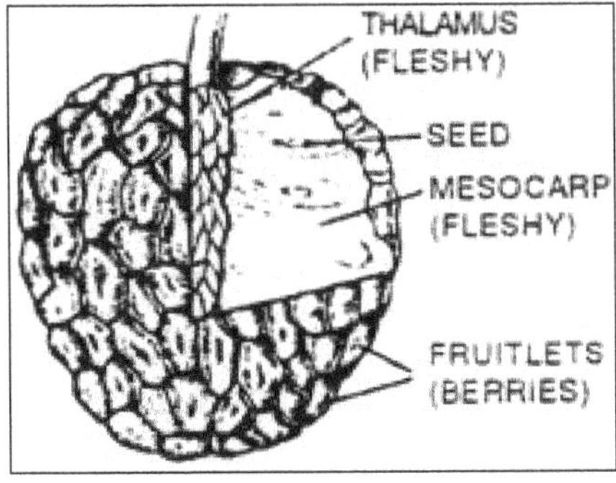

Etaerio of berries of custard apple.

Composite Fruits

A fruit developing from a complete inflorescence is called a multiple or a composite fruit.

There are two main types of composite fruits:

- Sorosis: This type of fruit is found in Mulberry, Pineapple and Jack fruit (kath-al). These fruits are derived from catkin, spike and spadix type of inflorescence.

  Mulberry (Morus indica) fruit develops from catkin in which fleshy perianth encloses dry achenes.

  In jack fruit, thick club-shaped peduncle has the flowers arranged on it. The fertile fruits have juicy, edible perianth lobes and the bracts form more or less juicy chaffs around them. The spines on the tough rind represent the stigmas of the carpels. Each seed is covered by a membranous testa. In Pineapple (Ananas sativus), the ovaries are not so conspicuous, edible portion being formed by peduncle, perianth and bracts. Each polygonal area on the surface represents a flower. This fruit develops from an intercalary spike.

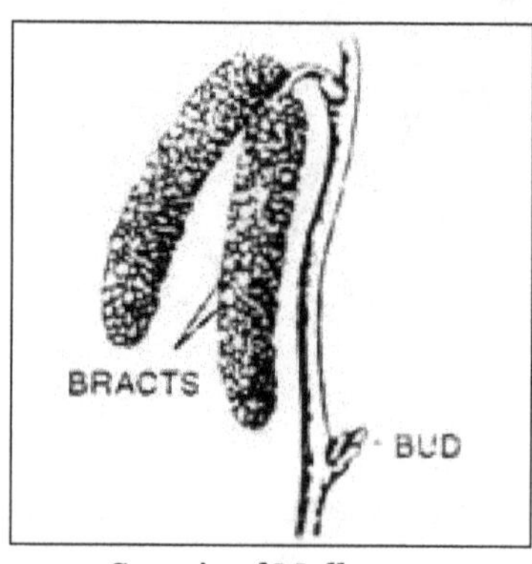

Sorosis of Mulberry.

- Syconus: This fruit develops from the hypanthodium type of inflorescence and is characteristic of Ficus. In figure, Banyan etc. Female flowers within the closed receptacle (which becomes fleshy) of the inflorescence develop into achenes giving rise to a multiple fruit of achenes.

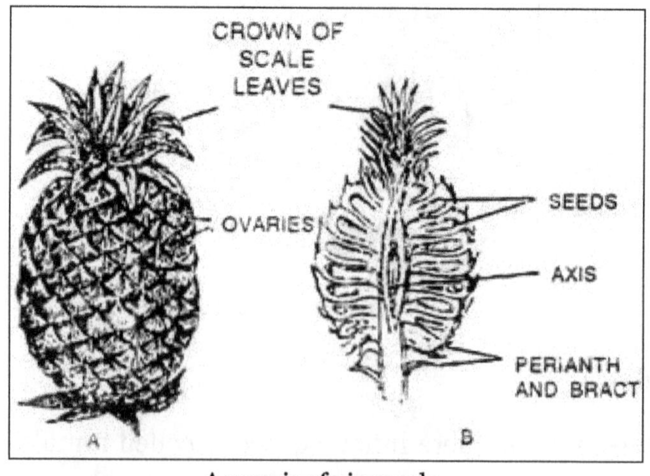

A sorosis of pineapple.

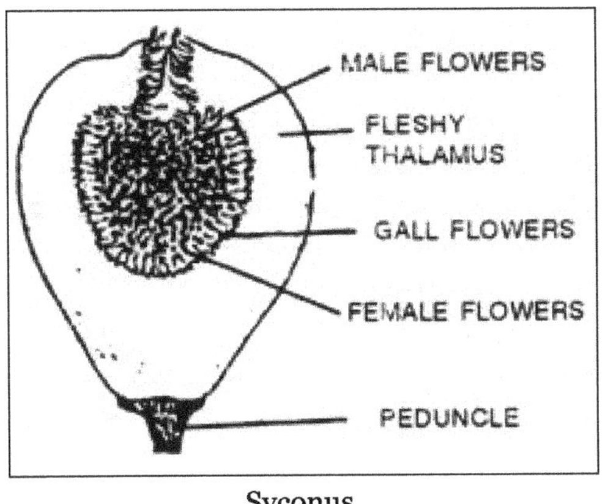

Syconus.

Function of Fruit

The function of the fruit in flowering plants is to protect the seeds and to facilitate their dispersal. Fleshy fruits are usually edible and are dispersed after being eaten. Color changes in fruit often signal ripeness and make ripe fruit easy to see. When fruits are eaten, the seeds pass through the gut mostly unaffected. In fact, in some cases, the partial breakdown of the seed coat stimulates germination. Excreted seeds are surrounded by nutrients (including minerals and simple organic compounds in the dung) along with a substrate for early growth. Of course, this strategy fails if the seeds are actually digested. As a plant defense mechanism, some plants produce chemicals that can make an animal ill if too many are digested. Some of the plants modified by man no longer have such chemical defenses.

Dry fruits are characteristic of seeds dispersed by the wind and other natural agents or animals. These fruits may have barbs, hooks, or a sticky surface that catches the coat of a passing animal and disperses the seeds. Winged fruits or those with tufts of hair are designed to catch the wind (for example, dandelions). Fruits of seaside plants often float and are resistant to water damage. The occurrence of coconut palms on seashores around the world is evidence of the success of this strategy. A fibrous husk traps air and conveys the hard seed to the high-water mark on a beach where the plant can become established. Some plants, particularly those of the California chapparal, thrive after fire. Fire-adapted fruits not only withstand fire temperatures, but may need them to trigger seed release.

Dry fruits often have specific adaptations for seed release. Many dry fruits are dehiscent, forming openings at specific locations in the wall. How the fruit opens determines whether seed release occurs slowly or all at once. Some dry fruits are indehiscent and do not open at all prior to seed germination.

Fruit Development and Ripening

The formation of fruits is typically triggered by sexual fertilization. Many changes occur in the flower accompanying fertilization, including the loss of petals, anthers, and stigma, modification or shedding of the sepals, development of the ovules into seeds, and formation of the embryo and endosperm within the seed. As part of fruit formation, the walls of the ovary surrounding the seeds are stimulated to resume cell division and to expand. The differences in the size and shapes of fruits are limited to some extent by the structure of the flower, and to an even larger degree by later patterns of growth.

Some fruits do not form seeds, such as bananas and seedless grapes. Fruits produced without seeds are examples of so-called parthenocarpic fruits, in which ovules (precursors of seeds) are formed but do not successfully fertilize.

Plant hormones play an important role in the formation and maturation of fruit. Fleshy fruits grow and thicken in response to hormonal growth signals emitted by fertilized seeds. In strawberries, for example, seed formation is highly successful except for the tip of the fruit, which is poorly developed. Where seeds are underdeveloped, so is the fruit. The stimulus for fruit production in this plant can be replaced by a plant hormone known as auxin, which is often produced by developing seeds. Fruit maturation and the development of fruit color are triggered by a later-occurring hormonal signal, produced by the gas ethylene. For grocery stores, fruit is often picked before becoming ripe because unripe fruit is not as easily bruised. To ripen the fruits for sale, a human-made gas related to ethylene is used after harvest, causing the immature fruit to develop its characteristic color and texture.

Once the fruit is ripe, the pedicel or stem that holds the fruit begins to seal itself off from the plant, under the influence of the hormone abscisic acid (ABA). When this hormone is produced, fruit drop is stimulated. To prevent fruit drop, another hormone called cytokinin can be used to inhibit the production of ABA and delay overripening and fruit drop. Oranges and other citrus crops can be harvested yearlong by inhibiting fruit drop and senescence through the application of a cytokinin. Citrus fruits, which normally mature in the winter, can thus be harvested year round.

A careful examination of the fruit reveals how the tissues change during development. In citrus fruits, like grapefruits and oranges, the bulk of the fleshy fruit is formed by small juice sacs, which originate from small hairs lining the inside of the pistil. These juice sacs are simply hairs that swell at different positions along their length, filling the fruit. The nature of these hairs can be seen by gently teasing a few sacs from the center of the fruit. The fleshiness of the tomato fruit is the result of the swelling of the placenta, a tissue that connects the seeds to the walls. Frequently, the ovary wall itself forms most of the fruit, but the exact region of thickening differs in each plant group. In squash, cucumbers, and pumpkins for example, the middle of the ovary wall grows thicker than the inner and outer layers, whereas in grapes, the inner wall grows thicker, and in watermelons, the outer wall is particularly thick. In dry fruits, the thickening of the ovary wall is sometimes accompanied by cell hardening, which is caused by chemical changes in the cell walls. These hardened cells form the walls of nuts and other hard fruits. In dry fruits, the walls of the fruit are no longer living.

Seed Anatomy

Seed is the characteristic reproductive body of both angiosperms (flowering plants) and gymnosperms (e.g., conifers, cycads, and ginkgos). Essentially, a seed consists of a miniature undeveloped plant (the embryo), which, alone or in the company of stored food for its early development after germination, is surrounded by a protective coat (the testa). Frequently small in size and making negligible demands upon their environment, seeds are eminently suited to perform a wide variety of functions the relationships of which are not always obvious: multiplication, perennation (surviving seasons of stress such as winter), dormancy (a state of arrested development), and dispersal. Pollination and the "seed habit" are considered the most important factors responsible for the overwhelming evolutionary success of the flowering plants, which number more than 300,000 species.

Fruit and seeds of the southern magnolia (Magnolia grandiflora). The fruit is an aggregate of follicles, and each follicle bears a single red seed.

The superiority of dispersal by means of seeds over the more primitive method involving single-celled spores, lies mainly in two factors: the stored reserve of nutrient material that gives the new generation an excellent growing start and the seed's multicellular structure. The latter factor provides ample opportunity for the development of adaptations for dispersal, such as plumes for wind dispersal, barbs, and others.

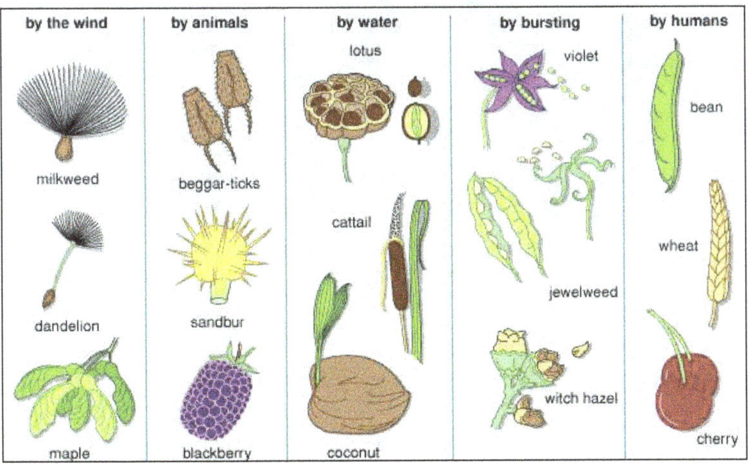

Economically, seeds are important primarily because they are sources of a variety of foods—for example, the cereal grains, such as wheat, rice, and corn (maize); the seeds of beans, peas, peanuts, soybeans, almonds, sunflowers, hazelnuts, walnuts, pecans, and Brazil nuts. Other useful products provided by seeds are abundant. Oils for cooking, margarine production, painting, and lubrication are available from the seeds of flax, rape, cotton, soybean, poppy, castor bean, coconut, sesame, safflower, sunflower, and various cereal grains. Essential oils are obtained from such sources as juniper "berries," used in gin manufacture. Stimulants are obtained from such sources as the seeds of coffee, kola, guarana, and cocoa. Spices—from mustard and nutmeg seeds; from the aril ("mace") covering the nutmeg seed; from the seeds and fruits of anise, cumin, caraway, dill, vanilla, black pepper, allspice, and others—form a large group of economic products.

Caraway seeds.

Castor bean seeds used to make oil cakes.

The Nature of Seeds

Angiosperm Seeds

In the typical flowering plant, or angiosperm, seeds are formed from bodies called ovules contained in the ovary, or basal part of the female plant structure, the pistil. The mature ovule contains in its central part a region called the nucellus that in turn contains an embryo sac with eight nuclei, each with one set of chromosomes (i.e., they are haploid nuclei). The two nuclei near the centre are referred to as polar nuclei; the egg cell, or oosphere, is situated near the micropylar ("open") end of the ovule.

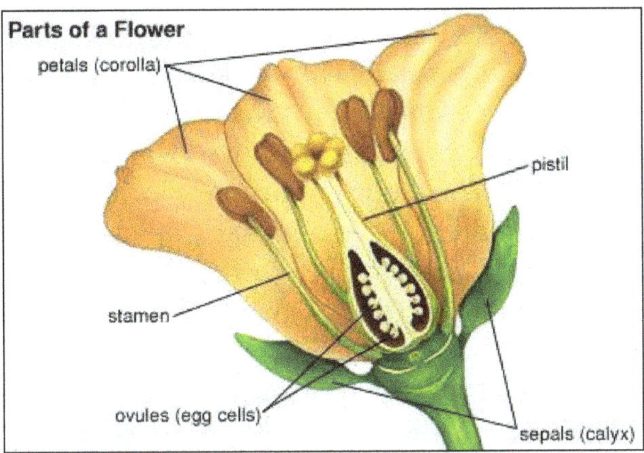

Diagram of a typical flowering plant (angiosperm).

With very few exceptions (e.g., the dandelion), development of the ovule into a seed is dependent upon fertilization, which in turn follows pollination. Pollen grains that land on the receptive upper surface (stigma) of the pistil will germinate, if they are of the same species, and produce pollen tubes, each of which grows down within the style (the upper part of the pistil) toward an ovule. The pollen tube has three haploid nuclei, one of them, the so-called vegetative, or tube, nucleus seems to direct the operations of the growing structure. The other two, the generative nuclei, can be thought of as non-motile sperm cells. After reaching an ovule and breaking out of the pollen tube tip, one generative nucleus unites with the egg cell to form a diploid zygote (i.e., a fertilized egg with two complete sets of chromosomes, one from each parent). The zygote undergoes a limited number of divisions and gives rise to an embryo. The other generative nucleus fuses with the two polar nuclei to produce a triploid (three sets of chromosomes) nucleus, which divides repeatedly before cell-wall formation occurs. This process gives rise to the triploid endosperm, a nutrient tissue that contains a variety of storage materials—such as starch, sugars, fats, proteins, hemicelluloses, and phytate (a phosphate reserve).

The events just described constitute what is called the double-fertilization process, one of the characteristic features of all flowering plants. In the orchids and in some other plants with minute seeds that contain no reserve materials, endosperm formation is completely suppressed. In other cases it is greatly reduced, but the reserve materials are present elsewhere—e.g., in the cotyledons, or seed leaves, of the embryo, as in beans, lettuce, and peanuts, or in a tissue derived from the nucellus, the perisperm, as in coffee. Other seeds, such as those of beets, contain both perisperm and endosperm. The seed coat, or testa, is derived from the one or two protective integuments of the ovule. The ovary, in the simplest case, develops into a fruit. In many plants, such as grasses and lettuce, the outer integument and ovary wall are completely fused, so seed and fruit form one entity; such seeds and fruits can logically be described together as "dispersal units," or diaspores. More often, however, the seeds are discrete units attached to the placenta on the inside of the fruit wall through a stalk, or funiculus.

The hilum of a liberated seed is a small scar marking its former place of attachment. The short ridge (raphe) that sometimes leads away from the hilum is formed by the fusion of seed stalk and testa. In many seeds, the micropyle of the ovule also persists as a small opening in the seed coat. The embryo, variously located in the seed, may be very small (as in buttercups) or may fill the seed almost completely (as in roses and plants of the mustard family). It consists of a root part, or radicle, a prospective shoot (plumule or epicotyl), one or more cotyledons (one or two in flowering plants, several in Pinus and other gymnosperms), and a hypocotyl, which is a region that connects radicle and plumule. A classification of seeds can be based on size and position of the embryo and on the proportion of embryo to storage tissue; the possession of either one or two cotyledons is considered crucial in recognizing two main groups of flowering plants, the monocotyledons and the eudicotyledons.

Seedlings, arising from embryos in the process of germination, are classified as epigeal (cotyledons aboveground, usually green and capable of photosynthesis) and hypogeal (cotyledons belowground). Particularly in the monocots, special absorbing organs may develop that mobilize the reserve materials and withdraw them from the endosperm; e.g., in grasses, the cotyledon has been modified into an enzyme-secreting scutellum ("shield") between embryo and endosperm.

Gymnosperm Seeds

In gymnosperms (plants with "naked seeds"—such as conifers, cycads, and ginkgo), the ovules are not enclosed in an ovary but lie exposed on leaflike structures, the megasporophylls. A long time span usually separates pollination and fertilization, and the ovules begin to develop into seeds long before fertilization has been accomplished; in some cases, in fact, fertilization does not occur until the ovules ("seeds") have been shed from the tree. In the European, or Scots, pine (Pinus sylvestris), for example, the female cones (essentially collections of megasporophylls) begin to develop in winter and are ready to receive pollen from the male cones in spring. During the first growing season, the pollen tube grows slowly through the nucellus, while within the ovule the megaspore nucleus, through a series of divisions, gives rise to a collection of some 2,000 nuclei, which are then individually enclosed by walls to form a structure called the female gametophyte or prothallus. At the micropylar end of the ovule, several archegonia (bottle-shaped female organs) develop, each containing an oosphere ("egg"). The pollen tube ultimately penetrates the neck of one of the archegonia. Not until the second growing season, however, does the nucleus of one of the male cells in the tube unite with the oosphere nucleus.

Although more than one archegonium may be fertilized, only one gives rise to a viable embryo. During the latter's development, part of the prothallus is broken down and used. The remainder, referred to as endosperm, surrounds the embryo; it is mobilized later, during germination of the seed, a process that occurs without delay when the seeds are liberated from the female cone during the third year after their initiation.

Pinecone and exposed seeds of the pinyon pine (Pinus edulis). Pinyon pines are gymnosperms and bear their edible seeds, known as pine nuts, in protective cones instead of fruit.

Form and Function

Seed Size

In the Late Carboniferous Period (about 315.2 million to 298.9 million years ago), some seed ferns produced large seeds (12 × 6 cm [5 × 2 inches] in Pachytesta incrassata). This primitive ancestral condition of large seeds is reflected in certain gymnosperms (Cycas circinalis, 5.5 × 4 cm [2.2 × 1.6 inches]; Araucaria bidwillii, 4.5 × 3.5 cm [1.8 × 1.4 inches]) and also in some tropical rainforest trees with nondormant water-rich seeds (Mora excelsa, 12 × 7 cm [4.7 × 2.8 inches]). The "double coconut" palm Lodoicea maldivica represents the extreme, with seeds weighing up to 27 kg (about 60 pounds). Herbaceous nontropical flowering plants usually have seeds weighing in the range of about 0.0001 to 0.01 gram. Within a given family (e.g., the pea family, Fabaceae), seed size may vary greatly; in others it is consistently large or small, justifying the recognition of "megaspermous" families (e.g., beech, nutmeg, palm, and soursop families) and "microspermous" ones (e.g., milkweed, daisy, heather, nettle, and willow families).

The smallest known seeds, devoid of food reserves, are found in orchids, mycoheterotrophs (nongreen plants that absorb nutrients from dead organic matter and live symbiotically with mycorrizal fungi—e.g., Indian pipe, Monotropa; coral root, Corallorhiza), carnivorous plants (sundews, pitcher plants), and total parasites (members of the families Rafflesiaceae and Orobanchaceae, or broomrapes, which have seeds weighing as little as 0.001 mg—about 3.5 hundred-millionths of an ounce). Clearly, seed size is related to lifestyle. Total parasites obtain food from their host, even in their early growth stages, and young orchids are mycoheterotrophs that receive assistance in absorbing nutrients from mycorrhizal fungi that are associated closely with their roots. In both cases only very small seeds that lack endosperm are produced. Dodders (Cuscuta) and mistletoes (Viscum, Phoradendron, Amyema) live independently when very young and accordingly have relatively large seeds.

Many plant species possess seeds of remarkably uniform size, useful as beads (e.g., Abrus precatorius) or units of weight—one carat of weight once corresponded with one seed of the carob tree, Ceratonia siliqua. In wheat and many other plants, average seed size does not depend on planting density, showing that seed size is under rather strict genetic control. This does not necessarily preclude significant variation among individual seeds; in peas, for example, the seeds occupying the central region of the pod are the largest, probably as the result of competition for nutrients between developing ovules on the placenta. Striking evolutionary changes in seed size, inadvertently created by humans, have occurred in the weed known as gold-of-pleasure (Camelina sativa), which grows in flax fields. The customary winnowing of flax seeds selects forms of C. sativa whose seeds are blown over the same distance as flax seeds in the operation, thus staying with their "models." Consequently, C. sativa seeds in the south of Russia now mimic the relatively thick, heavy seeds of the oil flax that is grown there, whereas in the northwest they resemble the flat, thin seeds of the predominant fibre flax.

Seed Size and Predation

Seeds form the main source of food for many birds, rodents, ants, and beetles. Harvester ants of the genus Veromessor, for example, exact a toll of about 15,000,000 seeds per acre (37,050,000 seeds per hectare) per year from the Sonoran Desert of the southwestern United States. In view of the enormous size range of the predators, which include minute weevil and bruchid-beetle larvae that attack the seeds internally, evolutionary "manipulation" of seed size by a plant species cannot in itself be effective in completely avoiding seed attack. With predation inescapable, however, it must be advantageous for a plant species to invest the total reproductive effort in a large number of very small units (seeds) rather than in a few big ones. The mean seed weight of those 13 species of Central American woody legumes vulnerable to bruchid attack is 0.26 gram (0.009 ounce). In contrast, the mean seed weight of the 23 species invulnerable by virtue of toxic seed constituents is 3 grams (0.1 ounce).

Seed Size and Germination

Ecologically, seed size is also important in the breaking of dormancy. Being small, a seed can only "sample" that part of the environment immediately adjacent to it, which is not necessarily representative of the generally prevailing conditions. For successful seedling establishment, there is clearly a risk in "venturing out" in adverse conditions. The development in seeds of mechanisms acting as "integrating rain gauges" should be considered in that light.

The Shape of Dispersal Units

Apart from the importance of shape as a factor in determining the mode of dispersal (e.g., wind dispersal of winged seeds, animal dispersal of spiny fruits), shape also counts when the seed or diaspore is seen as a landing device. The flatness of the enormous tropical Mora seeds prevents rolling and effectively restricts germination to the spot where they land. In contrast, Eusideroxylon zwageri does not grow on steep slopes, because its heavy fruits roll downhill. The grains of the grass Panicum turgidum, which have a flat and a round side, germinate much better when the flat rather than the convex side lies in contact with wet soil. In very small seeds, the importance of shape can be judged only by taking into account soil clod size and microtopography of the soils onto which they are dropped. The rounded seeds of cabbage species, for example, tend to roll into crevices, whereas the reticulate ones of lamb's quarters (Chenopodium album) often stay in the positions in which they first fall. Several seeds have appendages (awns, bristles) that promote germination by aiding in orientation and self-burial. In one study, for example, during a six-month period, awned grains of Danthonia penicillata gave rise to 12 times as many established seedlings as de-awned ones.

Polymorphism of Seeds and Fruits

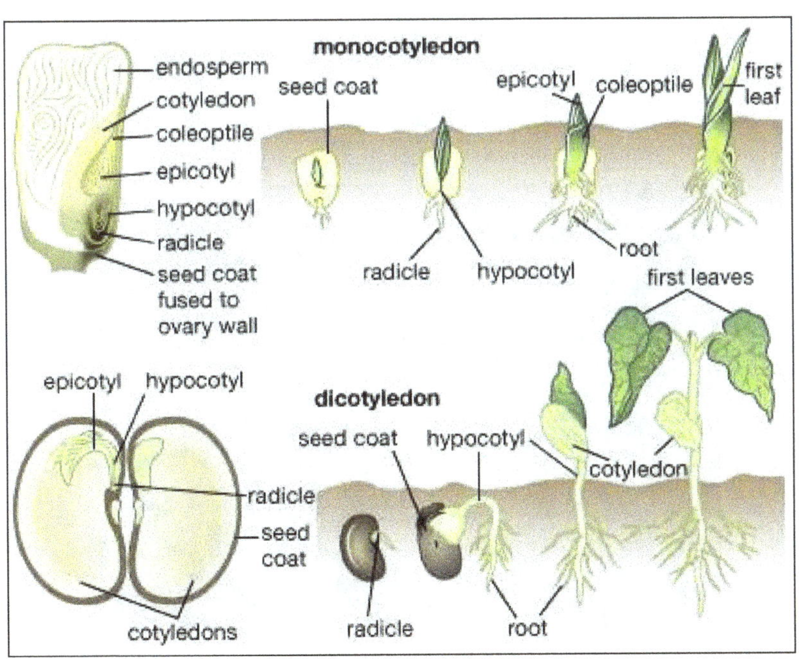

Some plant species produce two or more sharply defined types of seeds that differ in appearance, colour, shape, size, internal structure, or dormancy. In common spurry (Spergula arvensis), for example, the seed coat (part of the mother plant) may be either smooth or papillate (covered with tiny nipple-like projections). Here the phenomenon is genetically controlled by a single factor, so all the seeds of a given plant are either papillate or smooth. More common is somatic polymorphism, the production by individual plants of different seed types, or "morphs." Somatic polymorphism occurs regularly in saltbush (Atriplex) and goosefoot (Chenopodium), in which a single plant may produce both large brown seeds capable of immediate germination and small black ones with some innate dormancy. Somatic polymorphism may be controlled by the position of the two (or more) seed types within one inflorescence (flower cluster) or fruit, as in cocklebur, or it may result from environmental effects, as in Halogeton, in which imposition of long or short days leads to production of brown or black seeds, respectively. Since the different morphs in seed (and fruit) polymorphism usually have different dispersal mechanisms and dormancies, so germination is spread out both in space and in time, the phenomenon can be seen as an insurance against catastrophe.

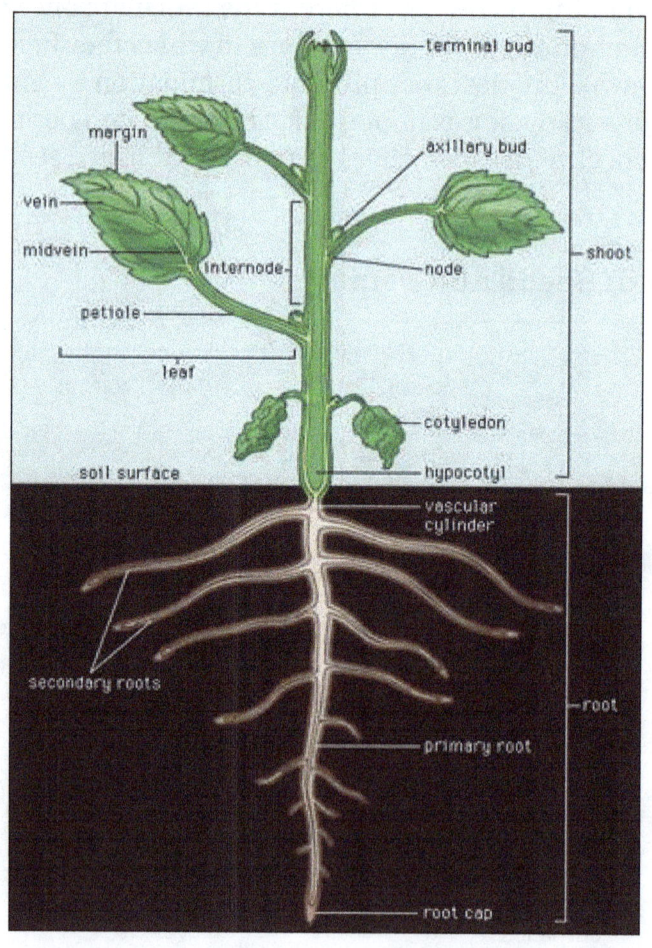

Cotyledon

Cotyledon from a Judas-tree (Cercis siliquastrum) seedling.

A cotyledon is a significant part of the embryo within the seed of a plant, and is defined as "the embryonic leaf in seed-bearing plants, one or more of which are the first to appear from a germinating seed." The number of cotyledons present is one characteristic used by botanists to classify the flowering plants (angiosperms). Species with one cotyledon are called monocotyledonous ("monocots"). Plants with two embryonic leaves are termed dicotyledonous ("dicots").

In the case of dicot seedlings whose cotyledons are photosynthetic, the cotyledons are functionally similar to leaves. However, true leaves and cotyledons are developmentally distinct. Cotyledons are formed during embryogenesis, along with the root and shoot meristems, and are therefore present in the seed prior to germination. True leaves, however, are formed post-embryonically (i.e. after germination) from the shoot apical meristem, which is responsible for generating subsequent aerial portions of the plant.

Comparison of a monocot and dicot sprouting. The visible part of the monocot plant (left) is actually the first true leaf produced from the meristem; the cotyledon itself remains within the seed.

The cotyledon of grasses and many other monocotyledons is a highly modified leaf composed of a scutellum and a coleoptile. The scutellum is a tissue within the seed that is specialized to absorb stored food from the adjacent endosperm. The coleoptile is a protective cap that covers the plumule (precursor to the stem and leaves of the plant).

Gymnosperm seedlings also have cotyledons, and these are often variable in number (multicotyledonous), with from 2 to 24 cotyledons forming a whorl at the top of the hypocotyl (the embryonic stem) surrounding the plumule. Within each species, there is often still some variation in cotyledon numbers, e.g. Monterey pine (Pinus radiata) seedlings have 5–9, and Jeffrey pine (Pinus jeffreyi) 7–13 but other species are more fixed, with e.g. Mediterranean cypress always having just two cotyledons. The highest number reported is for big-cone pinyon (Pinus maximartinezii), with 24.

The cotyledons may be ephemeral, lasting only days after emergence, or persistent, enduring at least a year on the plant. The cotyledons contain (or in the case of gymnosperms and monocotyledons, have access to) the stored food reserves of the seed. As these reserves are used up, the cotyledons may turn green and begin photosynthesis, or may wither as the first true leaves take over food production for the seedling.

Epigeal versus Hypogeal Development

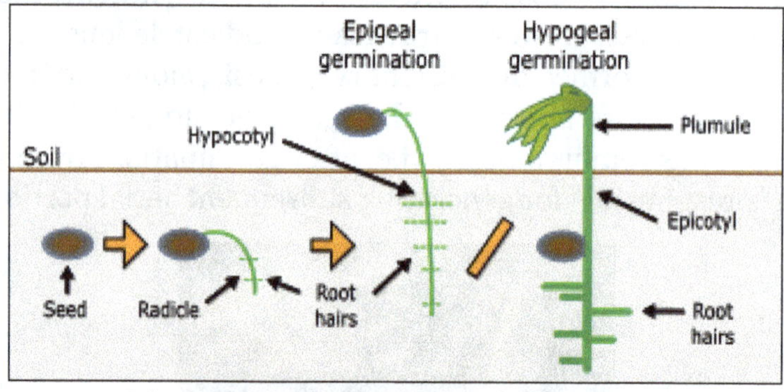

Schematic of epigeal vs hypogeal germination.

Cotyledons may be either epigeal, expanding on the germination of the seed, throwing off the seed shell, rising above the ground, and perhaps becoming photosynthetic; or hypogeal, not expanding, remaining below ground and not becoming photosynthetic. The latter is typically the case where the cotyledons act as a storage organ, as in many nuts and acorns.

Hypogeal plants have (on average) significantly larger seeds than epigeal ones. They are also capable of surviving if the seedling is clipped off, as meristem buds remain underground (with epigeal plants, the meristem is clipped off if the seedling is grazed). The tradeoff is whether the plant should produce a large number of small seeds, or a smaller number of seeds which are more likely to survive.

The ultimate development of the epigeal habit is represented by a few plants, mostly in the family Gesneriaceae in which the cotyledon persists for a lifetime. Such a plant is Streptocarpus wendlandii of South Africa in which one cotyledon grows to be up to 75 centimeters (2.5 feet) in length and up to 61 cm (two feet) in width (the largest cotyledon of any dicot, and exceeded only by Lodoicea). Adventitious flower clusters form along the midrib of the cotyledon. The second cotyledon is much smaller and ephemeral.

Related plants may show a mixture of hypogeal and epigeal development, even within the same plant family. Groups which contain both hypogeal and epigeal species include, for example, the Southern Hemisphere conifer family Araucariaceae, the pea family, Fabaceae, and the genus Lilium. The frequently garden grown common bean, Phaseolus vulgaris, is epigeal, while the closely related runner bean, Phaseolus coccineus, is hypogeal.

Peanut seeds split in half showing the embryos with cotyledons and primordial root.

References

- Fruit-and-its-types-explained-with-examples, plants-6398: biologydiscussion.com, Retrieved 05 April, 2020

- Kotpal, Tyagi, Bendre, & Pande. Concepts of Biology XI. Rastogi Publications, 2nd ed. New Delhi 2007. ISBN 8171338968. Fig. 38 Types of placentation, page 2-127

- Seed-plant-reproductive-part, science: britannica.com, Retrieved 16 May, 2020

- Stewart-Cox JA, Britton NF, Mogie M (August 2004). "Endosperm triploidy has a selective advantage during ongoing parental conflict by imprinting". Proceedings. Biological Sciences. 271 (1549): 1737–43. doi:10.1098/rspb.2004.2783. PMC 1691787. PMID 15306295

- 3-main-types-of-endosperm-development-in-flowering-plant, reproduction-in-plants-26787: yourarticlelibrary.com, Retrieved 19 June, 2020

INDEX